RAILWAYS
AND THE
LAKE
DISTRICT

DAVID JOY

PHOTOGRAPHY
GAVIN MORRISON

GREAT NORTHERN

Great Northern Books
PO Box 1380, Bradford, BD5 5FB
www.greatnorthernbooks.co.uk

ISBN: 978-1-914227-64-6

Design and layout: David Burrill

CIP Data
A catalogue for this book is available from the British Library

CONTENTS

PREFACE

Over forty years have elapsed since I first compiled a book featuring the outstanding railway photography of Gavin Morrison. In 1981 I was an editor at Dalesman Publishing, developing a list that in theory covered every conceivable aspect of the Lake District as well as the Yorkshire Dales but inevitably became slanted towards my obsession with railways.

I soon realised that Gavin had a skilled ability to portray steam in all its glory and to depict trains in the landscape rather than just shots of locomotives. This was especially the case when it came to the West Coast main line hugging the edge of the Lake District for a full forty miles including the magnificent Lune gorge and the heroic climb over Shap Fell.

Gavin went on to supply photos for numerous albums and books on railways in many parts of northern England. Unlike many photographers, he did not give up and put his cameras away when 'real' steam finished in 1968. Instead, he continued to capture the changing railway scene both at home and overseas. His collection is not only amazing in extent but is also held on a digital database. Gavin can usually find any photograph within a couple of minutes!

Frequently a harassed editor's salvation, he has continued to meet numerous challenging requests down the years. In 2022 we felt it was time to mark four decades of working together and take a radically different approach in a new book on railways and the Yorkshire Dales. Rather than my completing the text before selecting the photos, Gavin first chose his favourite images and I then wrote detailed captions for each picture along with introductory reading.

The resulting *Railways and the Dales* was well received. We felt it was a logical step to move onto the present *Railways and the Lake District*, looking not only at the main line over Shap but also the line round the coast offering magnificent views of both sea and mountains.

Branch lines are also included, although here we had to overcome a problem as Gavin had few photographs of two of the most significant. These are the pioneer Windermere branch, the only one still to remain open, and the much missed line from Penrith to Keswick and onto Cockermouth. It was the only railway to run through rather than merely into the Lake District.

Happily there was a solution to this difficulty. In 1984 I edited *Steam in the North West* featuring the work of the accomplished photographer Derek Cross. We corresponded for years on everything from cameras to farming and geology to classical music. At the end of it all he left me a large number of black & white prints including several of Lake District branches. I was delighted to have Gavin's agreement that a selection is included in these pages as they undoubtedly make the coverage more complete.

Overall our aim has been to capture Lake District railways in their immense variety from the late 1950s through to recent times. We feel this has been achieved.

David Joy,
January 2024

INTRODUCTION

In the space of a few hundred square miles, the Lake District contains England's highest mountains, greatest concentration of natural lakes and most dramatic scenery. Its exact boundary was not defined until as late as 1951 when it became England's largest National Park. The ultimate accolade had to wait for another 65 years when at the fourth attempt it was designated a UNESCO World Heritage Site.

The Lake District mountains adjoin the Pennine fells, creating a natural barrier which along with the tidal coastal estuaries severely retarded the region's progress. Daniel Defoe was awestruck by the 'unhospitable terror' of a land 'eminent only for being the wildest, most barren and frightful of any that I have passed over in England'. He referred to 'a chain of almost impassable mountains', seen in a totally different light when railway revolution struck Britain like a thunderbolt.

In 1830 the triumphant success of the world's first steam-hauled passenger railway between Liverpool and Manchester swiftly led to proposals for trunk main lines. It was a different age to that of today's HS2 and only eight years later it was possible to travel by rail not just from London to Birmingham but on to Manchester, Liverpool and Preston. A line as far as Lancaster was under construction. Stage coaches then had to be endured for an uncomfortable journey over Shap Fell and along the edge of the Lake District to Carlisle, where the first railway to be completed across England could be joined to reach Newcastle. These were heady times and there were already ambitious proposals not only to link Lancaster and Carlisle but to create an Anglo-Scottish main line by continuing north to Glasgow.

There was one immediate obstacle. It was the 'almost impassable' Lake District mountains and adjoining fells. Should a railway go round them or through them?

Favoured by a young and later eminent engineer Joseph Locke was a direct route from Lancaster through Kirkby Lonsdale and the Lune Gorge to Tebay. Ahead lay the bleak ridge of Shap Fell, which he proposed to pierce by a 1¼-mile tunnel before a gradual descent to Penrith and on to Carlisle. It was not at all to the liking of two towns, which feared being cast to one side and earnestly put forward rival schemes. Kendal, the largest settlement on the southern edge of the Lake District, wanted a line that would serve it by heading due north from Lancaster. It would then climb up Longsleddale, plunge into a 2¼-mile tunnel some 700ft deep under Gatescarth Pass and run alongside Haweswater to reach Penrith.

Far to the west there was much anguish in Whitehaven, utterly isolated and yet once the largest town in northern England after Newcastle and York. A heyday had seen it as a coal port busier than Bristol or Liverpool, but it now feared that a glorious past would quickly slip away. George Stephenson, the 'father of railways', was commissioned to help and came up with a radical proposal. A main line from Lancaster would skirt the Lake District mountains by cutting across Morecambe Bay on a causeway, span the Leven and Duddon estuaries and follow the coast to Whitehaven. Here it would link up with lines already being promoted locally to reach Carlisle. Although creating a route some thirty miles longer than the other two schemes, it would at no point be more than 40 feet above sea level. Trains using it would not be slowed down by steep gradients or wild winter weather as would be the case on the two inland routes.

Divisive argument could have dragged on for many a month had not the government in November 1839 taken the then unprecedented step of appointing a commision to review possible railway routes between England and Scotland. Priority was given to the major problem posed by the Lake District and commissions were not then an excuse for delay. Six months later a decisive report firmly rejected the coastal route owing to formidable construction challenges posed by a long causeway across Morecambe Bay. Engineering problems also told against a deep tunnel under Gatescarth Pass but this did not leave a direct Lancaster to Carlisle route as the only option. Instead there should be a railway that almost reached Kendal before making a dog's leg curve on acceptable gradients to join the route of the direct line at the entrance to the Lune Gorge. It was a true British compromise.

In a few carefully worded sentences the report paved the way for the whole future of railways and the Lake District. An Anglo-Scottish main line would skirt the edge of the mountains but would still pass within sight of them and only five miles distant from Ullswater. The hopes of Whitehaven were dashed and it would henceforth be served by a fragmented chain of purely local railways following the coast. There would be no spectacular trunk route penetrating the mountains to run alongside Haweswater and the heart of the Lake District would now have little more than a series of branch lines.

The commission's report on the route to be taken between Lancaster and Carlisle had one immediate outcome. Nothing happened. The initial splurge of railway construction in the mid-1830s died as speculators waited to see a return on their investments. There was a pause but then a sudden surge of money culminated in what became known as the Railway Mania. It became clear that there would be two main lines from London to Scotland – a West Coast route via Carlisle to Glasgow and a rival East Coast route through York and Newcastle to Edinburgh. By late 1843 they were effectively engaged in a race to see which would be the first to be completed.

The Lancaster & Carlisle Railway engaged Joseph Locke as its engineer-in-chief. Rather than tunnel under Shap Fell, he decided to save construction time by going straight over the top. Thus in haste were created the infamous four miles of 1 in 75, which were such a notorious handicap to West Coast operation for more than a century. Navvies working on the line rose to a peak figure of almost 10,000 and the mixture of English, Irish and Scots led to massive riots only quelled by military intervention. Despite near warfare, the 69 miles forming the largest single railway contract then placed were completed in 30 months and services began in December 1846. It was an amazing achievement. Opening of the Caledonian Railway meant the West Coast route reached Glasgow in February 1848, two years ahead of its East Coast rival which had to bridge the Tyne and Tweed estuaries.

An early development saw the end of local independence when the Lancaster & Carlisle Railway became part of the mighty London & North Western – dubbed the 'Premier Line'. It met the Caledonian at Carlisle, long known as the 'Border City'. Although less than twenty miles from the summit of Skiddaw and half that distance from the northern edge of today's National Park, Carlisle has always seemed a touch remote from the Lake District. Centuries of uncertainty over whether it was in England or Scotland may not have helped.

This proved no handicap to it becoming a unique meeting point of four English and three Scottish companies. Constantly at loggerheads, it was almost as if the various alliances felt obliged to feud in a city that had seen warfare since the days of Hadrian's Wall. One of the seven was the Midland Railway, which determinedly embarked on creating a third Anglo-Scottish route with the opening of its legendary Settle-Carlisle line in 1876. It was never totally successful and its attempts to penetrate the Lake District were rebuffed by the 'Premier Line'.

Instead it fostered tourist traffic to the area by working closely with the Furness Railway, which from tiny beginnings transformed Barrow from coastal hamlet to boom steel town. Unlike other local companies it ceased to be dependent on conveying minerals and was fortunate in having its own branches to Windermere and Coniston Water. It might have gone further had it not been for an increasingly powerful conservation lobby opposed to any further railways in the heart of the Lake District. No line was ever built to link Windermere with Keswick and proposals for branches into Borrowdale and Ennerdale got nowhere.

The relevant companies both large and small became part of the LMS in 1923. Apart from some closures in the slump of the 1930s, it made few fundamental differences and the main line over Shap remained remarkably little changed for another three decades. More by good luck than any forward planning, it provided a superb setting for railway photography through to the official end of steam in August 1968. Happily this occurred two years before completion of the parallel M6 motorway destroyed much of its essential character for ever. Then came electrification of the main line in 1974, but steam had not yet made its final appearance.

In the same year the historic counties of Cumberland, Westmorland and Lancashire North of the Sands were replaced by a single administrative county that took the even more ancient name of Cumbria. When British Railways finally relented and allowed a limited return of preserved steam, it led to 'heritage trains' as instanced by the Cumbrian Coast Express introduced in 1978 (see page 74). Its success was followed two years later by the Cumbrian Mountain Express, which has had a varied itinerary that frequently embraces steam haulage between Lancaster and Carlisle.

The spirit of the main line over Shap lives on and is fully covered in the first chapter of this book. A bizarre consequence of the nuclear age is that the chain of railways round the coast has survived in its entirety and these lines are described in chapter 2. Sadly, recognition of the value of preserved railways in National Parks came too late for the Lake District. Not only was the chance lost of a superb line centred on Keswick but the existing branch vanished in favour of a new trunk road. It is described in chapter 3 along with other lines that have suffered a similar demise. Finally, there is coverage of two locations where narrow-gauge steam can still be enjoyed in a magnificent setting that only the Lake District can offer. Not all is lost!

MAIN LINE OVER SHAP

The fascination and sheer variety of the Lancaster to Carlisle line is evident within the first few miles. It begins with the Lune viaduct, 55 feet above the river estuary and from the outset praised as 'a stupendous construction'. Within a few miles the West Coast route reaches Hest Bank – the only point where it is within sight of the sea – and there are glimpses across Morecambe Bay to the Lake District mountains. Next comes Carnforth, which became the junction for the line taking the long way round the coast to Carlisle via Whitehaven as described in the next chapter.

A steady 13-mile climb begins at Milnthorpe and soon reaches Oxenholme, junction for the branch to Kendal and Windermere (page 86). Here in steam days heavy trains often took on a second locomotive for the steepening gradient ahead. After an abrupt curve to the east the line comes out into the open on the shoulder of Hay Fell with views over Longsleddale, originally favoured as an alternative route through rather than round the Lake District to reach Carlisle. In the distance are high peaks such as Fairfield and Helvellyn.

The tracks soon curve back to the north and enter the most impressive stretch of mountain country on the English portion of the West Coast route. On the east side of the Lune gorge are the many peaks of the Howgills, which have been likened to sleeping elephants. To the west are the Borrowdale Fells, belatedly added to the Lake District National Park in 2018. Once a wonderfully peaceful place with bleating sheep and moorland birds providing the only sound betweeen trains, it all changed when the M6 motorway also came this way.

Tebay at the foot of the steep climb up Shap bank is the result of Joseph Locke's decision to save construction time and avoid a tunnel. Many northbound trains had to stop for banking assistance and the result was locomotive sheds with a supporting community that grew into some eighty houses. Railwaymen based there had such options as poaching in the River Lune or playing in their own brass band. Southbound trains passed through Tebay at great speed before picking up water at Dillicar troughs in the Lune gorge.

The four miles of Shap bank are scenically an anti-climax before the deep cutting close to the 937ft summit. It is then an easier descent down to Penrith, where the railway got off to a bad start by using part of the historic castle site for the station. Finally there is Carlisle with its impressive Citadel station, designed by the noted architect Sir William Tite and the nearest in Britain to a European-style railway frontier. Carlisle grew into one of the greatest of provincial railway centres with sprawling goods yards, huge engine sheds and all the related housing and industry that went with them.

A massive proportion of travellers between Lancaster and Carlisle have always had their eyes set firmly on Scotland. In the heyday of named trains it boasted expresses with such evocative titles as the Royal Scot and the Coronation Scot. Many would not even glance west towards the Lake District, but to a committed minority it was a favourite destination above all others. These services depended on the branch line from Oxenholme to Windermere as well as the railway heading deeply through the mountains from Penrith to Keswick and Cockermouth.

The LMS recognised the need by introducing the Lakes Express in 1927. It generally left Euston at noon and its through carriages had a complexity closely rivalling those of the better-known Atlantic Coast Express on the Southern Railway. A portion for Blackpool, and on Fridays for Southport, was detached at Wigan. The train would then divide at Lancaster, one half going right round the coast through Barrow and Workington to terminate at Maryport. Beyond Carnforth it called at the genteel resort of Grange-over-Sands and later stops included Ulverston and Foxfield, respectively enabling passengers to change for Lake Side (page 95) and Coniston (page 103).

The other half continued along the main line to Oxenholme where a further division took place, with one portion heading to Windermere complete with restaurant car – except on Fridays when this facility remained with the main train as far as Penrith. Here passengers who had travelled from south of Preston were instructed to notify the guard if they wished a stop to be made to set down at Troutbeck or Threlkeld. It must all have required a mastery of Bradshaw, if only to take on board that this portion of the Lakes Express would terminate at Workington a good half hour ahead of the town being reached by the Maryport carriages making their way round the coast.

The most surprising development in the pre-motoring age between the wars was the growth of Sunday services reaching Penrith from distant destinations and returning the same evening after passengers had hopefully enjoyed a good half-day in the Lake District. A variety of conveyances could take them over the five miles to Pooley Bridge, from where the 'steamers' *Lady of the Lake* and *Raven* plied the length of Ullswater. By July 1938 there were as

many as five return trains serving Penrith, and it is tempting to think that subtle differences reflected the occupations and preferences of those using them.

From Sunderland came a service calling at stations serving mining communities before reaching Barnard Castle via Durham and Bishop Auckland. It then headed over Stainmore to arrive in Penrith after almost three-and-a-half hours. Closely following was a train from Saltburn pausing at stations close to Middlesbrough and then Darlington. Little imagination is needed to visualise their respective complements of miners and steel workers, together with their families, plentifully equipped with more than adequate liquid refreshments to last until they got to journey's end. Any doubts as to where they were heading should have been put at rest by a large station nameboard 'Penrith for Ullswater Lake'.

Lugging bottles of beer onto the train was not a necessity for Newcastle citizens wanting a day away from Tyneside. Moreover they had a choice, with a Buffet Car Express for Blackpool leaving the city at 9.33am and calling only at Hexham and Haltwhistle prior to reversing at Carlisle to reach Penrith at 11.43am. Here passengers not hellbent on the seaside could leave the train and head for Ullswater before it returned from Blackpool in the evening. A second Buffet Car Express pulled out of Newcastle at 10.20am and at Penrith there was the option of continuing through to Keswick.

Most remarkable of all, and presumably catering for rather more than the average shipyard worker, was a Restaurant Car Express leaving Glasgow St Enoch at 9.55am. Travelling via the Glasgow & South Western main line, it called at Kilmarnock, Dumfries, Annan and Carlisle to be in Penrith by 1.10pm and then Keswick a shade before 2.0pm. A journey of four hours each way to have just an afternoon in the Lake District now seems an odd concept, although these were the glorious days of the restaurant car. Between late breakfast or early lunch outward and dinner on the return, it would be possible to adjourn for afternoon tea in the refined surroundings of the Keswick Hotel next to the station.

Penrith station staff must have found Sunday evenings challenging, as all five return trains left within little over an hour. First came the two from Keswick followed by the services to Sunderland and Saltburn, and then finally the return Buffet Car Express from Blackpool which did not reach Newcastle until 10.42pm. It was hopefully the end of a long and yet thoroughly enjoyable day out.

Sunday pleasantries on this scale never returned after the Second World War and both Penrith and Ullswater remained relatively quiet until the age of mass motoring took hold in the late 1950s. There were few fundamental changes on the West Coast main line until the demise of steam in 1968 when Shap had become a favourite haunt of photographers. The diesel age ended the need for a permanent fleet of banking engines at Tebay, which quickly declined into a ghost community until rescued by an M6 service station. The real transformation came with electrification through to Glasgow in 1974. Shap ceased to be an obstacle and drivers soon got used

to topping the summit at almost 100mph before having to brake for the ensuing speed restriction.

By this time the Lakes Express had made its last run as a named train in August 1964. Of the branches it served, only that to Windermere now remains and it is a shadow of its former self. Reducing the line to a single-track siding as part of the financial case for West Coast route electrification created the bizarre situation of excursions having to terminate at Oxenholme and disgorge their passengers onto droves of road coaches while the trains were worked empty stock over the 50 miles to Carlisle for servicing. As a sop the junction station was renamed 'Oxenholme Lake District' in 1987.

Thirty years after the last train to Keswick, a similar marketing ploy created the new name of 'Penrith North Lakes' in 2003. Passengers determined to have an afternoon on Ullswater can still travel south from Glasgow or from Newcastle by changing at Carlisle. Here there is at least a chance to admire the Victorian Gothic architecture of Citadel station. Once a byword for confusion and delay, it is possible to recall the days when those suffering food at its main refreshment room puzzled over the fireplace with its Latin motto. In a way not intended, they fully agreed when it was explained that 'Faciam ut Hujus Loci Semper Memeris' loosely translated as 'I will cause you ever to remember this place'!

Above

Class 87 No. 87019 *Sir Winston Churchill* is on the rear of an afternoon service from Glasgow Central to London Euston calling at Lancaster on 1st August 2001. In common with all stations on the Lancaster & Carlisle Railway, it was designed by the distinguished architect Sir William Tite. There were at first few changes when the company became a small part of the mighty London & North Western Railway in 1859, but by late Victorian times the town council was protesting about 'the danger and inconvenience caused to passengers by reason of the totally inadequate accommodation'. A rebuild was finally completed in 1902 when the station largely assumed its present form. Scarcely intrusive is the overhead wiring for the 1974 electrification of the West Coast main line.

Above

Far more prominent is the catenary on the connecting line operated by the Midland Railway to serve both the resort of Morecambe and the port of Heysham. Highly experimental when introduced in 1908, it had alternating current at 6,600 volts and came to be used as a test bed for later schemes on Britsh Railways before abandonment in 1966. An EMU has just arrived at Lancaster on 9th October 1965.

Below

A major feature immediately north of Lancaster station is the impressive Lune viaduct, built of timber when opened in 1846 and largely replaced with wrought-iron spans after only twenty years. Crossing on 20th January 2003 is a Virgin Voyager that had left Penzance at 7.20am and still had a long journey ahead to Glasgow. Passenger comfort by this date was giving way to distinctly cramped accommodation on such lengthy journeys.

Above

The Metrovick Co-Bos were among the least successful of all diesel locomotives introduced under the BR modernisation plan in the late 1950s. It was unfortunate for the Lake District that they handled many of its services, although No. D5705 is managing to accelerate away from Morecambe South Junction with a Workington to Euston express on 29th August 1964.

Opposite, above

The variety of traffic seen on the same day at at this point included 9F 2-10-0 No. 92006 on a northcound car train.

Oppposite, centre

In the opposite direction it was soon followed by Stanier Mogul No. 42961 joining the West Coast main line with a train comprising vans of many kinds. It is using a curve belatedly provided by the London & North Western Railway to compete with the Midland for Morecambe and Heysham traffic.

Opposite, below

Hest Bank is the point where the main line comes closest to the coast at Morecambe Bay and the sea can be seen on a full tide. 'Princess Royal' No. 46206 *Princess Marie Louise* provides an impressive sight as it hurries north with what is thought to be a Birmingham to Glasgow express on 7th June 1960.

Opposite, above

The changing scene at Hest Bank, where Stanier 2-6-4T No. 42613 has called with an up local service on 7th June 1960. At the extreme left is the bay platform once used by Morecambe trains and also visible are camping coaches, which in the 1930s became fashionable for 'basic holidays' in several selected locations. Here they provided users with a magnificent view across the Bay to the Lake District fells.

Opposite, below

Electrification has taken place by 21st May 2007 but the local traffic is being handled by a 'Pacer' in the bright yellow livery of Merseyrail.

Above

The steam heyday is still surviving on 7th June 1960 as wonderfully clean 'Princess Coronation' No. 46238 *City of Carlisle* heads the up Mid-Day Scot past Hest Bank. Small wonder that this was a perfect location for young trainspotters such as the author of this book, who would happily await one express after another as their parents searched for what was surely less exciting marine life on the sea shore.

Above

Steam was in abrupt decline only four years after the photo on the previous page of a clean 'Princess Coronation' Pacific hauling a top-link named express. On 26th August 1964, sister locomotive No. 46225 *Duchess of Gloucester* has been reduced to humble duty on a parcels train. It had only another three weeks left in service and was withdrawn having completed over 1½ million miles during its 26-year career. Here the 'Duchess' is passing Carnforth No. 1 signal box, a large and lengthy London & North Western Railway structure closed in 1973 as part of the main line resignalling.

Below

On the same date, shunting duties are busily taking place in the goods yard at Carnforth. Originally no more than a wayside station, it quickly grew into an important junction once Barrow had been connected with the West Coast main line as described in the next chapter. Engaged in this unglamorous but essential task is 'Jinty' No. 47662. A total of 422 such locomotives were built between 1922 and 1930 by six different companies, with this example being withdrawn in January 1966.

Above

Low-lying coastal surroundings are soon left behind at Carnforth and from Milnthorpe the line is on an unbroken climb for the next thirteen miles. By Oxenholme it is on a shelf above Kendal and the Kent valley, seen at top left as 'Britannia' No. 70024 *Vulcan* heads south with a van train on 6th April 1963. Surprisingly still surviving are the tall London & North Western lower-quadrant signals, positioned at this height to aid visibility on a curve.

Lower

Change is clearly coming on 6th August 1966, two years before the official end of steam on British Railways. 'Britannia' No. 70028 *Royal Star* is in far from pristine condition as it approaches Oxenholme with a northbound summer Saturday extra. Waiting in the loop are 8Fs Nos. 48352 and 48471 on a long train providing rail for a main-line upgrade. At extreme right is the engine shed, which would soon be demolished but was once busy providing extra locomotives for the steeper gradient ahead.

Above

In the first of two photos of Oxenholme station taken on 6th August 1966, an up express is departing behind Class 47 No. D1859 (later renumbered 47209).

Below

Shortly afterwards there was a clear path for a class DMU to follow it with a service that has come off the Windermere branch, featured in chapter 3. Passengers using the branch-line trains then had the unusual luxury of an overall roof, although the station as a whole had a poor reputation. Accounts of the time refer to a dismal subway, depressing waiting rooms and a generally discouraging atmosphere. Renaming as 'Oxenholme Lake District' in 1987 was accompanied by much improvement, and the general muddle on the right of this picture is now a much needed car park for a new generation of passengers.

Above

Once through Oxenholme the true drama of the main line over Shap begins in earnest as it starts its dog's-leg curve up to Grayrigg summit. It emerges from a series of cuttings and comes out into the open on the shoulder of Hay Fell. On a clear day there are distant views across to many famous summits among the Lake District mountains, but closer at hand are the folds and gentle curves of the Howgills. Snow adds a flourishing touch on 5th January 2003 as the 9.36am from Glasgow Central to Euston is about to cross Docker viaduct. In Virgin livery is class 87 No. 87015 *Howard of Effingham*, named after the commander of English forces during the 1588 battles against the Spanish armarda.

Above

The 13-mile climb from Milnthorpe to Oxenholme gets steeper on the remaining eight miles to Grayrigg summit, much of it at 1 in 106. This was too demanding for heavy freights, especially in the last years of steam traction when locomotives were in a sorry state. Typical of the period on 1st April 1967 is 'Britannia' No. 70032 with its *Tennyson* nameplates already removed. It is working hard at Lambrigg on a lengthy goods train with a banking engine in the rear. In common with all other surviving members of the class, it had been withdrawn by the end of the year.

Below

Three years earlier on 26th September 1964, sister locomotive No. 70041 is still carrying *Sir John Moore* nameplates but otherwise its appearance is characteristic of express motive power that had seen better days. It has reached Grayrigg summit with a northbound freight. The steam above the middle of the train is probably a banking engine about to return to Oxenholme.

Above

The centre portion of the 75ft high six-arch Docker viaduct spanning a tributary of the River Mint. Self-evident is the high standard of construction with limestone used for the piers and facings of the arches. When inspecting the Lancaster & Carlisle Railway the government inspector commented that he had 'never seen handsomer or more substantial masonry on any other railway'. Class 60 No. 60047 in Coal Sector livery is crossing with a train of oil tanks from Dalston to Stanlow on 14th May 1994.

Above

Still in undercoat livery, brand new class 47 No. D1624 is heading a test train on 26th September 1964. It was renumbered 47043 in 1974 and then 47566 in 1980. The derelict platforms are those of Grayrigg station, closed ten years before this photograph was taken.

Opposite, above

A Crewe to Glasgow summer Saturday extra approaches Low Gill on 18th August 1962 behind 'Jubilee' No. 45671 *Prince Rupert*. It was withdrawn the following year after covering almost 1½ million miles since new in 1935.

Opposite, below

Once the M6 had been opened in 1970, the isolation of the West Coast main line was over less than two miles from Grayrigg. Clearly visible in this scene near Beckfoot is the oblique bridge carrying the motorway, which then runs close to the railway all the way to Carlisle. On 13th May 1989 class 86 No. 86414 is heading the northbound Cumbrian Mountain Express, a 'heritage train' that is often steam-worked.

Above

Goods trains running at their low speed with a motley assortment of wagons had their advantages. After photographing 'Britannia' No. 70032 *Tennyson* at Lambrigg on 1st April 1967 (page 22), there was ample time to drive into the Lune gorge and secure this second shot. There will be a further delay at Tebay to take on a banker for the climb up Shap. Soaring in a complex pattern of gentle folds and lovely curves are the Howgills with a series of summits culminating in the 2,220ft peak of The Calf. Amazingly unknown and unvisited at the time of this photo, they gained full National Park status in 2015.

Opposite, above

The same year saw the Borrowdale Fells on the west side of the railway added to the Lake District National Park. This picture is taken from their foothills but could not be repeated today as the M6 now runs through the relatively flat land close to the line. Steam is already losing its supremacy on 18th August 1962 when class 40 No. D233 *Empress of England* is in charge of a Birmingham to Edinburgh express.

Opposite, below

The Howgills remain a dominant feature on 13th May 1989 as a Freightliner leaves them behind at Beckfoot on its way to Lancaster. It is double-headed by class 86 No. 86413 and class 87 No. 87012, named *Royal Bank of Scotland* from 1988 until 1998 when it became *Coeur de Lion*.

Opposite, above and below

Glorious isolation and not another sign of life on 18th August 1962 as 'Princess Coronation' No. 46252 *City of Leicester* pounds through the Lune gorge in charge of a mere five carriages. This is the summer-only Lakes Express, which has detached the rear portion of its train at Oxenholme to be taken on to Windermere. The Pacific is then continuing with the front portion to Penrith where it will come off and these carriages will be taken over the branch line to Keswick and Workington by a much smaller locomotive.

Above

The same location on the same day sees 'Royal Scot' class No. 46133 *The Green Howards* heading a Manchester to Glasgow express. It is getting perilously close to steam's final fling on such services and this locomotive was withdrawn a year later. Here it is in good stride on the stretch of line known as the 'Dillicar Straight' – it is the only portion not snaking its way through the Lune gorge on a series of gentle curves. It also had a functional purpose as shown on page 32.

These three photographs are all taken from a similar vantage point in the Lune gorge looking across more open country towards Shap Fell. Prominent on the left is Jeffrey's Mount.

Above

On 28th November 1964 there is a good covering of snow on higher ground as 'Clan' No. 72007 *Clan Mackintosh* heads into the gorge with an up freight. The course of the line can clearly be followed as it curves round towards Tebay.

Opposite, above

A much lower camera position has to be adopted on 2nd May 2007 as uncomfortably close to the left there is now the M6. A Freightliner from Coatbridge to Basford Hall is in charge of Nos. 86627 and 86614 in contrasting liveries.

Opposite, below

On the same day, the Freightliner is followed by a Royal Mail train from Shieldmuir to Warrington. It comprises three four-car sets Nos. 325001, 325005 and 325015, introduced in 1995 and designed to carry bulk mail at speeds of up to 100mph.

Opposite

A footplate permit from Preston to Carlisle on 17th May 1961 provides a rare opportunity for photography from the cab. 'Jubilee' No. 45642 *Boscawan* will have its tender scoop lowered as it picks up water from Dillicar troughs. These were a now forgotten feature of the 'Dillicar Straight', which had the essential virtue of being dead level. Apart from limited use in France and the USA, water troughs were confined to Britain. They revolutionised long-distance travel, enabling expresses to travel for hundreds of miles without stopping. For sheer spectacle, Dillicar troughs were without parallel in Britain. Northbound expresses leaving the gorge were building up the maximum amount of speed for the steep climb up Shap bank that lay ahead. Smoke, steam and spray merged with the background of the mountains to create an almost mystical effect.

Above

Southbound expresses thundering down from Shap at high speed hit the troughs like an arrow in full flight. It needed skilled work by the fireman to avoid picking up too much water, which would overflow from the tender onto the front carriage. Passengers sitting with the toplight windows left open could easily be soaked! This will not be a problem as 'Princess Coronation' No. 46255 *City of Hereford* heads towards the troughs with a Perth to Euston express on 7th May 1960. It looks as if ancient tradition is being maintained with a horse box at the head of the train. In the background is Tebay station and on the left is the loop used to stable freight trains waiting for banking asistance up to Shap summit.

Opposite, above

5MT No. 45290 passes Tebay on 26th May 1958 and starts the climb of Shap bank with its four miles at 1 in 75. With a six-coach summer Saturday extra from Manchester to Glasgow it certainly has no need for banking assistance. The terrace houses on the hillside form part of the railway community that developed to cater for the engine sheds and the crews of banking engines. Their numbers increased when Tebay became the junction for a line trailing in on the centre left of this picture. Closed in 1962, it was built to bring coke from County Durham to Barrow (see chapter 2).

Opposite, below

Much harder work lay ahead for 'Princess Royal' No. 46212 *Duchess of Kent*, which has run through Tebay without stopping for a banker. On Sunday 12th June 1959 it is heading the Royal Scot instead of the normal 'Princess Coronation'. Built in 1935, withdrawal took place in October 1961 after running almost 1½ million miles.

Above

The M6 can be seen in the distance but the sound of class 37s Nos. 37172 and 37150 will have clearly been audible long before they come in sight at Greenholme. On 5th July 1984 they are in charge of the exceptionally heavy Clitheroe to Uddingston cement train.

Above

Passenger trains were frequently double-headed up the bank. In another scene at Greenholme a summer Saturday afternoon train from Manchester to Glasgow has 5MT No. 44769 assisting 'Jubilee' No. 45726 *Repulse* on 6th August 1960.

Opposite, above

Shap has rarely been troubled by deep snow but the cold winter air on 28th November 1964 gives a crisp look to 5MT No. 44762 near Shap Wells. The van train, including a Travelling Post Office, will have a banking engine at the rear.

Opposite, below

A later train at the same location on the same day has Fairburn 2-6-4T No. 42210 as its banker. It was withdrawn in April 1967, eight months before the official end of steam over Shap when Tebay shed was closed.

Opposite

The probably unique sight on 7th May 1960 of a North Eastern class J21 No. 65033 climbing past Scout Green signal box with a rail-tour that had reached Tebay via Stainmore summit. It is only a three-coach train and hence is unaided. Withdrawn in April 1962 as the last of its class, the 0-6-0 remained in Darlington works for some four years before being preserved. A distinctive feature in this picture is the exceptionally tall London & North Western lower-quadrant signal.

Above

In the silence that descended on Scout Green in pre-motorway days, most trains could be heard a long way off coming slowly up the bank and there was ample time for a photographer to get ready. There was one exception. The Lakes Express had dropped off a good half of its train at Oxenholme and a 'Princess Coronation' with the few remaining carriages would often surmount the climb at close on 60mph. The crew would be anxious to get home and end their shift at Penrith or Carlisle. Looking in superb form on 6th August 1960 is No. 46252 *City of Leicester*.

Above

Shap summit, 916ft above sea level. Ivatt 4MT No. 43009 is setting off with the daily pick-up goods to Tebay on 24th August 1963. The signal box in the background could be busy day and night, as all banking engines had to be crossed over and sent back in readiness for their next duty.

Opposite, above

An up express headed by class 87 No. 87010 *King Arthur* has just passed the summit on 20th July 1974. It is the year when the line was newly electrified. Drystone walls instead of timber fencing form an attractive feature but there are few signs of any tree cover.

Opposite, below

A fraction over 30 years makes all the difference. Trees both deciduous and coniferous make the summit seem much less bleak. Class 390 'Pendolino' No. 390032 is forming a Glasgow to Euston express on 6th September 2005.

Above

It is early days for English Electric Type 4s on West Coast route expresses as No. D211, later named *Mauretania*, passes Shap station with a Perth to Euston service on 29th August 1959. Over two miles on the Penrith side of the summit, the station closed in 1968.

Opposite

After the drama of the Lune gorge and Shap bank, there is inevitably a sense of anti-climax on the descent to Penrith. Yet there are points of interest as shown by these three photographs all taken at Harrison's Sidings, serving a lime works about two miles north of Shap station.

Above

On 19th May 1960 the morning express from Glasgow Central to Birmingham comprises as many as 17 carriages. 'Princess Royal' No. 46205 *Princess Victoria* makes an awesome noise as it struggles south at about 30mph.

Centre

Taken from the signal box steps, 'Patriot' No. 45510 is both unrebuilt and unnamed. On 6th August 1960 it is drifting down the bank towards Penrith with a van train from Crewe.

Below

A pair of 5MTs Nos. 44942 and 45148 head a summer Saturday relief to the up Royal Scot on 6th August 1961. They are making considerably faster progress than the main train shown on page 46.

North of Harrison's Sidings, the line runs through the little-known valley of the River Leith. The photographs on these two pages are taken at Thrimby Grange, where trees straddling the railway make for attractive surroundings that are all too short-lived.

Opposite, above

Maroon-liveried 'Princess Coronation' No. 46248 *City of Leeds* looks magnificent against the woodland as its pounds south with a Glasgow to Manchester train on 11th August 1962. This would normally be a class 40 diesel working, and it is probably acting as a substitute for a locomotive that has failed at Carlisle.

Opposite, below

In more recent times class 60 No. 60008 is approaching the woods on 3rd May 2007. The lengthy train of empties from Redcar has almost reached its destination at Hardendale Quarry, a supplier of limestone to steelworks on Teeside.

Above

On the same day in 2007, fell country is still prominent in the distance as a four-car Virgin Voyager heads south.

Above

Another view of 'Princess Coronation' *City of Leeds* at Thrimby Grange, this time sporting a fine headboard as it takes the up Royal Scot past a loop line on 6th August 1960.

Below

The same location three days earlier with 'Clan' No. 72004 *Clan Macdonald* in filthy external condition on a summer Saturday extra. Built in 1952, it was withdrawn in December 1962 after little more than ten years in service.

Above

Another locomotive looking terribly unkempt is 'Britannia' No 70016, originally named *Ariel*. Surviving until August 1967, it is taking an up freight past the typically London & North Western signal box at Clifton & Lowther on 8th May 1965.

Below

Far more presentable on 30th August 1958 is 'Princess Coronation' No. 46253 *City of St. Albans*, passing through Penrith with the morning Glasgow Central to Birmingham service. The tracks on the left converge to form the line through to Keswick and Workington (page 112).

Above

A Birmingham to Glasgow Central express drifts into Carlisle on 9th August 1960 behind 'Scot' No. 46134 *The Cheshire Regiment*, withdrawn in December 1962 after covering over two million miles. There is still open countryside behind the train, but the vast array of lines on the left indicates that a startling change in the surroundings is imminent.

Below

Moments later, incoming expresses pass the full extent of Carlisle Upperby engine shed with dozens of locomotives in steam. Prominent on 22nd May 1961 are 'Jubilee' No. 45700 *Amethyst* and 'Princess Coronation' No. 46248 *City of Leeds*.

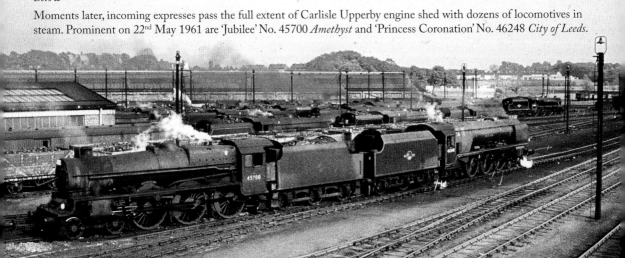

Above

'Princess Coronation' No. 46230 *Duchess of Buccleuch* passes Upperby shed on heading out of Carlisle. It is 9th August 1960 and the express is the up Mid-Day Scot. A quick sprint had been required to obtain this photograph, as the same train was earlier photographed running into Carlisle station from Glasgow (see page 53).

Below

A final glimpse of Upperby shed on 1st September 1963 with the coaling tower in the left background. In the centre is 'Princess Coronation' No. 46256 *Sir William A. Stanier, F.R.S.*, named after the Chief Mechanical Engineer responsible for introduction of the class in 1937. Behind the coal wagons is sister locomotive No. 46235 *City of Birmingham*.

Above

The historic Carlisle London Road where the city's railway age began on 19th July 1836. Here gathered a crowd estimated at 40,000 strong to witness the opening of the western end of the Newcastle & Carlisle Railway – the first line across England. It became part of the North Eastern Railway in 1862 when its passenger services finally connected with those on the West Coast main line. The London Road terminus then became a goods depot. Passing it on 17th May 1980 is 'Peak' No. 45019 with a service that will head south over the Settle-Carlisle railway to Nottingham.

Opposite, above

When the Midland completed its legendary Settle-Carlisle railway in 1876, it shared tracks into the city with the existing North Eastern line from Newcastle. Immediately before the junction at Petteril Bridge it had its own engine shed and goods sidings at Durranhill, of which few traces now remain. Passing the site on 13th February 1982 is 'West Country' No. 34092 *City of Wells*, making an impressive departure from Carlisle with the Cumbrian Mountain Express. This 'heritage train' often journeyed north to Carlisle over the West Coast main line as shown on page 25.

Below

The complexities of Carlisle. Approaching Carlisle station on 14th September 1985 is class 85 No. 85002 at the head of the Sealink service from Euston to Stranraer, where it will connect with ferries to Larne in Northern Ireland. Passing under the bridge on the right is the line from Petteril Bridge Junction into the station carrying services from both Newcastle and Leeds via the Settle-Carlisle railway. The lines on the left, electrified to take West Coast main-line freight traffic clear of the passenger station, suffered a disaster the previous year. A runaway goods train travelling at excessive speed derailed and caused so much damage that the cost of rebuilding could not be justified. Official closure took place in December 1985.

Above

Carlisle Citadel station, where the Lancaster & Carlisle Railway met the Caledonian Railway to continue the West Coast main line through Scotland to Glasgow. Closure of the goods avoiding line in 1985 meant all freight traffic had to pass through the station, causing congestion at peak periods and especially at night. Entering the station on 2nd May 2002 with an up Freightliner is the unique class 86/5 No. 86501, modified to give better power and adhesion. Although performance appeared to have been improved, no other members of the class were similarly treated. The Scottish-looking buildings in the background are close to West Walls where the Caledonian initially had its goods yard. The constant smoke and noise caused intense friction between the company's Presbyterian directors and the Dean of Carlisle Cathedral, whose study overlooked the railway.

Above

Carlisle Citadel in steam days. The up Mid-Day Scot is arriving on 9th August 1960 behind 'Princess Coronation' No. 46230 *Duchess of Buccleuch*. This locomotive achieved a remarkable record in being allocated to the same engine shed – Glasgow Polmadie – from 1940 until withdrawal in 1963.

Below

On the same day, class 5MT No. 45455 runs into Citadel station on a service that has come from Dumfries on the former Glasgow & South Western Railway. This company joined the Caledonian at Gretna Junction, the two rival concerns having an uneasy relationship in Carlisle that was to last for their entire existence.

Opposite, above

New era at Citadel station on 11th September 2012 with Pendolinos Nos. 390141 and 390118 waiting to leave with departures at 12.38 and 12.40.

Opposite, below

A Virgin Voyager departs beneath the overall roof in 2002 with a Cross Country working to the south. In the bay platform is a blue-liveried DMU used on Newcastle services. Above it on the station wall are a series of impressive murals.

Above

Three of the murals photographed in August 2021. From left to right, the locomotives are 'Princess Royal' No. 46203 *Princess Margaret Rose*, 5MT No. 45013 and 'Jubilee' No. 45724 *Warspite*. They are depicted standing outside the cathedral-like end screens of Citadel station, removed in 1958 and replaced by corrugated steel and plain glass when the overall roof was substantially reduced in area.

Above

The age when Carlisle Citadel served the trains of seven companies may have long gone, but with one major exception all the lines they used remain open. There is still ample variety in the liveries of local services as shown by the three photographs on these pages. On 17th March 2001 class 158 No. 158902 in West Yorkshire red livery is forming a Glasgow to Leeds service. It is standing beneath the substantial footbridge spanning four tracks in the centre of the station.

Opposite, above

In the same location, class 143 No. 143622 in Tyne & Wear livery is waiting to depart to Dumfries on 30th September 1990.

Opposite, below

A sad loss in 1969 was closure of the Waverley route from Carlisle through remote Border country to Hawick and onto Edinburgh. The bay it used at the north end of Citadel was still known as the 'Waverley platform' on 17th March 2001, when it is occupied by class 156 No. 156504 sporting the livery commonly seen on services to Glasgow via Dumfries.

Above

Dark winter days and wet platforms provide a completely different atmosphere at Citadel on 5th December 1994. Waiting to depart with a main-line train to the south is class 47 No. 47746.

Opposite, above

On 17th March 2001 an afternoon service from Edinburgh to Birmingham arrives behind class 86 No. 86240. Its name *Bishop Eric Treacy* is especially significant in the annals of railway photography. Known as the 'Railway Bishop', he was acclaimed for his outstanding camera work, which included many pictures taken on his beloved main line over Shap. In 1973 he opened the preserved Lakeside & Haverthwaite Railway (see page 96). Eric Treacy died in 1978 leaving an archive of 12,000 photos to the National Railway Museum.

Opposite, below

Engine change on 30th September 1990. Class 47 No. 47463 has arrived with an up express and waiting to take it south is class 86 No. 86406.

Above

Class 153 No. 153358 provides solitary one-coach accommodation on a Whitehaven service on 14th May 2002. It is in the bay platform normally allocated for workings on the coast line. On the far side of the station the sun shines on 'Arriva' DMUs used on Newcastle services.

Below

A Corus Rail train comes off the Whitehaven line on 17th March 2001 behind class 60 No. 60006. Originally named *Great Gable*, many Lake District devotees were dismayed when it was renamed *Scunthorpe Ironmaster* in 1997. There is much more about the line round the coast from Carnforth through Whitehaven to Carlisle in the chapter that starts on the page opposite.

2

ROUND THE COAST

Whitehaven was devastated when government commissioners took the decisive step in 1840 of rejecting a main line to Scotland that would pass through the town. The 12,000 inhabitants could only shrug their shoulders and make the most of their life in total isolation. Miners continued to hack rich seams of coal extending far out to sea and the port was constantly crammed with ships taking vast tonnages over to Ireland.

Only ten miles distant was Ennerdale Water, even today the least visited of all the major lakes, but it was then a life of unremitting toil. A glorious lake surrounded by high mountains might as well have been on another planet.

There may have been a crumb of comfort that an attempt to expand Maryport, a few miles up the coast, was faltering. The Maryport & Carlisle Railway had been formed to build a link with the 'Border City' and thus connect with the existing line across England from Newcastle. Money was running out and it ultimately took eight years to complete a railway only 28 miles long in six separate stages. Once Wigton was reached it was at least possible for trains to be met by 'large and commodious' coaches, which headed over to Keswick and took visitors to the northern heart of the Lake District.

It was 1845 before the entire line was at last opened as the first part of what became 'the great way round the coast'. It was now the height of the Railway Mania and the imminent opening of the Lancaster to Carlisle main line must have rubbed salt into the wound when it came to Whitehaven's isolation. The town was finally goaded into promoting a connecting line to Maryport, completed in 1847.

Prospects for a railway south from Whitehaven to join the West Coast main line near Lancaster were deeply discouraging. The only source of traffic in the entire 75 miles was the small town of Ulverston and the three major estuaries of the Duddon, Leven and Kent formed major handicaps. Small wonder that scant attention was paid to the most remote corner of all. On the tip of the Furness peninsula what was little more than a glorified mineral tramway was being built to carry slate and ironstone down to jetties at a small hamlet with less than a hundred inhabitants. There was no concept that it was destined for an astounding future. Its full name was Barrow-in-Furness.

A first stage was formation of the Furness Railway, which with great difficulty brought four steam locomotives from Liverpool across Morecambe Bay on the deck of a ship. A seven-mile 'main line' was completed from Barrow to slate quarries at Kirkby, together with a branch to ironstone mines at Dalton, and then by 1848 had been extended to Broughton to handle traffic from the flourishing Coniston copper mines. This at last provided the stimulus to complete a connecting line south from Whitehaven along the coast, but even the fastest train of the day took 2¼ hours to cover the 35 miles to Broughton, from where the Furness took passengers onto a deep-water pier at Piel, close to Barrow. They would then board a steamer to Fleetwood and hopefully catch onward connections by rail to Manchester and Liverpool.

The first faint stirrings of tourist traffic to the Lake District that had begun at Wigton were now increased when the same steamers brought Blackpool holidaymakers across the Bay. From Piel pier the Furness Railway took them to Dalton where they boarded coaches onto Newby Bridge for a sail on Windermere. Up at Whitehaven a new station hotel overlooking the coal dust and chaos of the port had the temerity to offer 'extensive sea views' from apartments ideal for the 'nobility and families visiting the Lakes'. It failed, but more successful was another hotel at St Bees that arranged visits to Ennerdale Water and the truly spectacular Wastwater.

A key step saw the Dalton branch extended to Ulverston, even though it was then a fashionable inland resort and did not relish the coming of the railway. There now remained the greatest challenge of all, which was to build viaducts across the Leven and Kent estuaries to connect with Carnforth and hence reach Lancaster. Their completion in 1857 gave Whitehaven reasonably direct access to the rest of England and its fortunes ought to have been on the eve of transformation. It was discovered that the rich hematite iron ores on the fells close to the town were perfect for the new Bessemer process of steel-making. So too were those in Furness centred around Dalton. What happened next was due to two aristocratic landowners who had nothing in common apart from their wealth.

The whole of Whitehaven was owned by the 2nd Earl of Lonsdale, noted as a collector of fine furniture and opera singers to relieve his lifelong bachelorhood. He rarely visited the town and showed little enthusiasm for its railways. Even though scarcely using his seat at Whitehaven Castle, he still insisted that the nearby line should be concealed from view in a long tunnel. The new age of steel-making should have been seized as the perfect opportunity to create a co-ordinated network of railways to handle the lucrative ore traffic. Instead, it was left to separate companies and Whitehaven stagnated amid constant petty squabbling. The field was left wide open for

the all-conquering London & North Western Railway to take over the line extending northwards to Maryport past ironworks at Workington.

It was a totally different picture on the Furness Railway, which from the outset had been promoted by the 2nd Earl of Burlington with his seat at Holker Hall close to the Leven estuary. On inheriting the richest dukedom in England and becoming the 6th Duke of Devonshire, he had virtually unlimited resources to develop his fiefdom centred on Barrow. The former hamlet rapidly grew into a town with a population of 18,000, its own iron and steelworks and the development of massive docks intended to rival those of London and Liverpool.

Various mineral railways were completed to extract the now precious iron ore, but a more fundamental change came in 1866. It was possible to absorb the line built by the Earl of Lonsdale to connect Whitehaven with the Furness Railway at Broughton. Owing to lack of any commitment it had become a dilapidated single-track with just three timetabled trains a day often carrying as few as five passengers. It was a case of the tail wagging the dog, as within twenty years the Furness had successfully expanded from a short tramway into a 75-mile system linking Whitehaven and Barrow with the West Coast main line.

Doubts about the wisdom of the new acquisition were quickly cast to one side. Close to the point where the line swooped round the west side of the Duddon estuary was a station serving the then tiny village of Holborn Hill. Here at nearby Hodbarrow had been discovered what proved to be the world's then largest deposit of hematite iron ore. The result within the space of a few years was the creation of Millom with 4,000 inhabitants and the Furness Railway was able to serve a second town to grow out of the tiniest beginnings.

By the early 1870s the affairs of the Furness

were at their zenith and it was unique among railway companies in retaining feudal control exercised by two dukes and a lord. Yet there were already signs that the boom time would not last. New processes of steel-making ended dependence on hematite and there was soon a slump with soup kitchens on the streets of Barrow. Grandiose plans for massive docks were never completed, but at least the Furness had the determination to survive.

Hitherto minerals rather than passengers had been by far the dominant traffic and little thought had been given to the tourist potential of a line running for much of its length along the coast and skirting the Lake District.

The position now changed, although not always with immediate success. Of all places, it was remote Seascale north of the Esk estuary that was chosen to develop a large resort. With supreme optimism it was hailed as a pending 'Eastbourne of the North' and was to have a grand hotel, promenades and numerous villas. Only a few crescents and shops were ever completed, as it soon became obvious that it was just too far from any major centre of population.

It was a happier picture at temperate Grange-over-Sands, on the edge of Morecambe Bay and less than ten miles from Windermere. It was also much closer to Leeds and Bradford. The Furness Railway made a more cautious attempt to emulate the Cornish Riviera by laying out ornamental gardens and constructing a promenade close to the station, which was rebuilt in a style matching that of a nearby hotel offering every luxury.

The real success came in the golden age before the First World War with the revival of steamer services across Morecambe Bay from Fleetwood to Barrow, where Ramsden Dock station had replaced Piel Pier. As related in the next chapter, connecting trains were now able to take passengers up branch lines for tours of the Lake District and cruises on Windermere and Coniston Water.

The steamers were not resumed after the war, but the Furness Railway managed to retain its independence until the 1923 Grouping. So too did the Maryport & Carlisle, the first part of the coastal chain of railways to be completed back in 1845. Both concerns now became tiny cogs in the LMS, the world's largest transport organisation and a huge centralised bureaucracy. Positive developments on the coast line in the 1930s were few and far between, and British Railways showed a lack of enthusiasm to develop the region.

Through services from Whitehaven to London ended in 1966 to be replaced by DMUs either terminating at Barrow or going no further than Lancaster or Preston. The future was beginning to look bleak and the least used section of line between Whitehaven and Barrow might well have closed. Such a possibility was cast aside following establishment of the world's first nuclear power station at Sellafield and the associated highly controversial transport by rail of irradiated fuel elements.

Recent years have seen much improved services out of Barrow, many of them working through to Manchester Airport. Yet essentially the line hugging the coast remains a backwater, which for those not in a hurry offers a great way round and a fascinating detour of some four hours over the 130 miles between Lancaster and Carlisle. Although built in stages based first on Maryport, then Whitehaven and finally Barrow, many travellers making their initial journey over the line are likely to begin at Carnforth and hence the photographs in this chapter follow the same order. For much of the distance there are glorious views of both the sea and distant mountains, and between Silecroft and Drigg the line runs through the portion of the Lake District National Park that extends down to the coast. On a good day it can be truly magnificent.

CARNFORTH TO BARROW

Above

Carnforth achieved fame in war-weary 1945 when it became the locale for the most celebrated of railway films, David Lean's *Brief Encounter*. The station still looked weary when class 31 No. 31201 came off the Barrow line with a permanent-way train on 22 January 1995. An excellent restoration has since embraced a heritage centre where visitors can have a longer enounter with the film and its local links. On the right in this view is the West Coast main line, where attempts to reopen the through platforms closed in 1970 have proved unsuccessful.

Opposite, above

Carnforth also soon became noted for Steamtown when its former engine shed began a new role as a 'live steam museum'. One of its most impressive displays took place on 31st May 1975. Lined up from left to right are French Pacific 231K22, 4079 *Pendennis Castle*, 4771 *Green Arrow*, B1 61306, German Pacific 011104 and 92220 *Evening Star* – the last steam locomotive to be built by British Railways.

Below

The matchless atmosphere of a steam locomotive on shed as seen at Carnforth on 26th August 1964. 'Britannia' No. 70030 was named *William Wilberforce* after the leader of the movement to abolish the slave trade. There were local associations, as some small ports on nearby Morecambe Bay had clandestine links with slavery.

Above

The Furness Railway was able to develop several small coastal resorts when collapse of its mineral traffic forced a switch to tourism. Closest to holidaymakers from mill towns in Lancashire and Yorkshire's West Riding was Arnside, which was previously a tiny fishing village with some twenty dwellings. Class 156 No. 156428 has crossed the Kent estuary on 5th December 1994 and is arriving with one of the through services from Barrow to Manchester Airport introduced earlier that year.

Opposite, above

On the opposite side of the Kent estuary is Grange-over-Sands, which similarly grew into a popular but genteel seaside town on the edge of Morecambe Bay. 'Jubilee' No. 5690 *Leander* is departing towards Carnforth with a rail-tour on 5th September 1979. On the seaward side is the promenade built by the railway.

Opposite, below

Grange-over-Sands station, built in 1865 to replace an earlier structrure, has been exquisitely restored with its glass canopies, ironwork columns and ornamental brackets all in white, red and green. Class 156 No. 156469 is calling with a Manchester service on 5th March 2018.

Above

Cark & Cartmel station long saw regular use by the 6th Duke of Devonshire, who had his seat at nearby Holker Hall and was fundamental in the development of both the Furness Railway and the town of Barrow. Class 31 No. 31119 is passing with an afternoon service from Manchester on 23rd June 1990.

Opposite, above

On the same date, an earlier Manchester to Barrow service is heading across the 48 spans of Leven viaduct, a major undertaking completed in 1857. A violent gale blowing up the estuary in 1903 overturned a ten-coach train – fortunately without loss of life – and resulted in provision of a wind pressure gauge to prevent similar disasters in extreme weather.

Opposite, centre

A higher tide in the Leven estuary on 25th April 1998 when the viaduct is being crossed by the Pathfinder 'Barrow Buoy' special from Bristol. With a pair of class 56s at each end, the head locomotive at this stage was No. 56114.

Opposite, below

Plumpton Junction, where a short branch once extended to the left of this photograph and terminated near Conishead Priory Hydro. Class 31 No. 31432 is heading a Manchester train on 2nd October 1993. On the skyline is the Hoad Monument commemorating Ulverston-born Sir John Barrow, founding member of the Royal Geographical Society, and designed to resemble an Eddystone lighthouse.

Above and opposite, below

Ulverston station looking towards Barrow on 23rd June 1990. Built in 1872-4, it symbolises the Furness Railway's pinnacle of achievement with its Italianate flamboyancy of scale, tall clocktower complete with corner urns and richly monogrammed ironwork. The unusual arrangement of the island platform was designed to enable rapid transfer from westbound trains to services to Windermere Lake Side (see page 95). 'Pacer' No. 142010 is heading towards Barrow with a local service.

Above

A rail-tour at Ulverston on 5th May 1976 behind the unusual combination of *Flying Scotsman* piloted by ex London & North Western Railway No. 790 *Hardwicke*. Five days earlier the same two locomotives had played a prominent part in the centenary celebrations of the Settle-Carlisle railway.

Above

The Furness Railway effectively created Barrow as its own town with massive iron and steelworks as well as docks optimistically intended to rival those of London and Liverpool. Its brashly designed Central station, reflecting a more prosperous age, was destroyed in 1941 air raids and its replacement has few distinguishing features. Class 156 No. 156421 is leaving with a service to Manchester Airport on 25th April 1998.

Opposite, above

The same day saw one of the more interesting specials to the town in the shape of the Pathfinder 'Barrow Buoy' from Bristol, seen earlier in these photographs on page 69. 'Top and tailed' by class 56s, this view shows No. 56114 complete with headboard

Opposite, below

Class 56 No. 56029 is at the head of the special as it leaves Barrow's Ramsden Dock, thus providing a rare use of track by a passenger train. The dock originally catered for paddle steamers bringing holidaymakers from Fleetwood to the Lake District as well as sailings to Belfast. One of the few remaining pieces of the port's once vast railway network, it owes its continued existence to handling controversial nuclear fuel shipped in for transit to Sellafield. *Pacific Pintail*, known locally as a 'nuclear ship', is berthed in the dock.

Above

Between Barrow and Whitehaven is the least used portion of the coast route with a distinct character all its own. A favourite location for photographers is Eskmeals viaduct on the stretch of coast where the Lake District National Park extends down to the sea. A pair of immaculate Direct Rail Services class 20s, Nos. 20301 and 20302, fresh from royal train duty, are returning from Workington to Bristol with a Pathfinder tour on 1st June 1996.

Opposite, below

The viaduct is crossed on 2nd September 1980 by the Cumbrian Coast Express, a 'heritage train' successfully inaugurated two years earlier. It is headed by No. 850 Lord Nelson, well away from its haunts in former Southern Railway days.

Above

Midland Compound No. 1000 and 'Jubilee' No. 5690 *Leander* form an impressive pair of locomotives with a fine exhaust as they power northbound with the Cumbrian Coast Express on 5th May 1980. They are rounding the curve just north of Millom, where the town seen in the background shows scant evidence that it owes its origins to an ironworks handling vast deposits of iron ore prior to closure in 1968.

Above

Small stations still retaining many of their original features include Bootle, which is also in the portion of the National Park that embraces the coast. Unfortunately there are no signs of passengers as 'Pacer' No. 142004 in the distinctive livery of Greater Manchester Council calls on 1st June 1996. Apart from Askam and Millom, all intermediate stations between Barrow and Whitehaven were in 1977 reduced to the status of request stops.

Below

Foxfield, junction for the now closed branch to Coniston (page 103), has kept its 1909 signal box and attached waiting shelter as well as an early water tower and the station master's house. 'Pacer' No. 142035 is heading north on 5th December 1994.

Above

In the days of regular freight on the coast line, class 40 No. 40094 approaches Seascale with a lengthy cement train on 8th May 1976. There is little to suggest that this remote location was unwisely chosen by the Furness Railway in a failed attempt to create a major resort.

Below

There was again a shortage of passengers when class 153 No. 153363 called at Ravenglass on 1st June 1996 and there was no need for the wooden steps to help access from the low-level platform. Other days can be hectic as this is now the base of the preserved Ravenglass & Eskdale Railway (page 117), which uses the goods shed in the centre of this photo as its workshop and the main station building on the left as its own pub – the Ratty Arms.

Above

It is only too obvious when an unspoilt twenty-mile stretch of coastline extending north from Millom and the isolated peak of Black Combe comes to an abrupt end. 'Pacer' No. 142056 is entering totally different surroundings on 26th August 1998 as it arrives at Sellafield, base of British Nuclear Fuels. Always controversial, local criticism is tempered by the fact that it employs over 4,000 and has been a major factor in retention of a railway between Barrow and Whitehaven.

Opposite, above

Just a small part of the Sellafield complex on the same date. In the foreground are class 20 No. 20310 and class 37 No. 37611, two of the large fleet built up by Direct Rail Services (DRS) following its 1995 formation by British Nuclear Fuels. Originally confining operations to traffic between Barrow's Ramsden Dock and Sellafield, it expanded rapidly as the UK's only authorised carrier of irradiated material by rail and diversified to the extent of operating luxury passenger trains.

Opposite, below

Shrouded in secrecy and known locally as 'hush-hush' workings, a pair of DRS class 37s Nos. 37218 and 37604 are heading south past Maryport with a nuclear flask bound for Sellafield on 12th April 2011.

Above

Class 156 No. 156469 has still a long way to go on 10th October 2010 as it leaves St Bees with a service from Carlisle to Preston via Whitehaven and Barrow. Welcome signs are the floral hanging baskets on the red sandstone walls of the station building and leisure usage through the handling of cycle traffic.

Left

Earlier that day another class 156 No. 156438 is entering the passing loop at St Bees. This arrangement results from singling of the line beween Sellafield and Whitehaven as an economy measure.

Above

Class 25 No. 25300 at Whitehaven Bransty on 2nd September 1980. It has just emerged from the ¾-mile tunnel built at the insistence of the 2nd Earl of Lonsdale so that he could be spared sight of the railway on his very occasional visits to Whitehaven Castle.

Right

Dereliction at Parton, north of Whitehaven, became symbolic of the former mining era. Far below a class 108 DMU is heading along the coast on a stretch of line prone to damage by stormy seas.

Above

Workington station looking past its prime on 6th December 1994 as 'Pacer' No. 142056 prepares to depart with a service from Carlisle to Barrow. Prior to acute depression in the 1930s, the town also had Central station on a competing line created by local ironmasters.

Below

Workington Port is now one of the largest in north-west England and has a rail terminal fully equipped for handling container traffic. Class 66 No. 66164 is preparing to depart with a loaded train to Carlisle on 12th April 2011.

Above

Disaster struck Workington in mid-November 2009 when freak rainfall in the Lake District led to the River Derwent sweeping away all the town's road bridges. Direct Rail Services received high praise for its immediate response. Within ten days it was operating a 'Floodex' shuttle service across the surviving rail bridge to a temporary station at Workington North, here seen busy handling traffic on 21st April 2010.

Right

Signboard at the temporary station, which remained in use until 28th May 2010.

Above

The coast line continues to hug the coast before turning inland at Maryport. On clear days there are extensive views across the Solway Firth to the Scottish fells near Dumfries. A couple of class 53 units Nos. 153315 and 153363 are working a Barrow to Carlisle service on 21st April 2010.

Below

On the same day, there is both yellow gorse and blue sea to raise the spirits of passengers on a service approaching Maryport. It started at Preston at 8.30am before reaching Lancaser and then taking a journey of some four hours right round the coast to Carlisle.

Above

The 12.00 freight from Workington Port to Carlisle, shown on page 82, approaches Maryport on its way to Carlisle. Class 66 No. 66164 remains at the head of the lengthy train of UBC containers.

Below

One of the emergency services provided by DRS following the Workington flood disaster leaves Maryport on 21st April 2010. Like most such workings it was 'top-and-tailed' and here has class 47 No. 47501 closest to the camera. On the platform behind the rear locomotive is a 'bus shelter' – all that remains on the site of the once massive station buildings and offices of the proudly independent Maryport & Carlisle Railway.

3

LAKE DISTRICT BRANCH LINES

THE WINDERMERE BRANCH

The dog's-leg curve taken by the West Coast main line in order to come as close as possible to Kendal, the largest settlement between Lancaster and Carlisle, did not entirely satisfy the town. Rather than build a short connecting link, it was decided to take the bold step of promoting a branch that would continue to the shores of Windermere close to Ambleside. It would foster the town's trade and at the same time become one of the first railways to develop tourism.

The poet Wordsworth was incensed when the proposals were announced in August 1844. His work had done much to encourage early travellers to the Lake District, but he now cast railways in the role of creeping blight and feared that what he saw as hallowed ground would swiftly be overrun by visitors. He wrote letters to the most influential newspapers and people of the day, including Gladstone, claiming that the railway would destroy the beauty of the country and be 'highly injurious to its morals'. In October he penned his famous sonnet, beginning 'Is there no nook of English ground secure from rash assault?'

Wordsworth's protests were way before effective environmental campaigning. At the age of 74, he was out of step with prevailing Victorian ideals of expansionism and free trade. He was accused in the national press of unrivalled snobbery and attacked for his 'spirit of exclusiveness'. His initial support received from affected landowners was dissipated when it was announced that the line was to be cut back to end at Birthwaite, near Bowness and

well above the lake. As if to stress that elitist protest had achieved little, the terminus immediately took the name Windermere. It came to be known unofficially as Windermere Town to avoid confusiion with the later Lake Side (page 95).

The impressively large station initially existed in glorious isolation, forming the railway gateway to much of the Lake District. From here connecting coach services departed to Ambleside, Coniston, Hawkshead, Keswick and Cockermouth. The only nearby building was the Windermere Hotel, constructed to coincide with completion of the line and linked to the terminus by a private drive.

The railway soon attracted increasing numbers of long-stay holidaymakers and also opened up the immediate area to people at two very different ends of the social spectrum. The wealthy, especially 'cotton kings' from Manchester, were able to use Windermere as a weekend and summer refuge, building large houses in even larger grounds which offered both splendid views across the lake and isolated their occupants from a different type of visitor arriving by train. This was the working-class day-tripper, who in the time available was unable to venture far from the station and was compelled to jostle with crowds thronging the promenades on the lake shore at Bowness. The inaugural excursion to Windermere ran in August 1847 and in its first full year the line carried 120,000 passengers, two-thirds of them between May and October.

Within a decade there was a building boom and the fears of Wordsworth had come to pass. Harriet Martineau, writing her *Complete Guide to the English Lakes* in 1855, noted: 'A few years previously the area was so secluded that it was some distinction even for the most travelled man to have seen it. Now there is a Windermere railway station, and a Windermere post office and hotel – a thriving village and a populous locality.' The settlement soon grew into a small town and there were fears that its surroundings were becoming 'overcrowded with villas'. Excursion traffic increased dramatically and Whit Monday could see throngs approaching ten thousand descend on the terminus.

At the opening it had been prophesied by Kendal's mayor that the day would come when express trains from Manchester to Windermere would 'bring the merchant and manufacturer after he had finished his correspondence or left his Exchange, down to the shore of that beautiful lake, and if he likes carry him back next morning in time for breakfast'. By the early twentieth century this vision had largely been realised. In 1910 the Windermere Express was covering the journey between the two centres in just five minutes over two hours. As an indulgence, the cotton barons living close to the lake were provided with an exclusive club carriage fitted with comfortable armchairs, writing desks, sherry cabinets and a big clock. It was available only to holders of first-class season tickets. They had to apply to the club committee for admission, which was by election and involved payment of a hefty annual subscription.

In terms of incoming tourist traffic it was recognised that the Lake District with its fame resting on the minority appeal of scenic splendour and literary associations could never compete with the nearby Lancashire coastal resorts. Determined efforts were nevertheless made and by the 1930s there was the Lakes Express from London augmented by through services from Manchester and Liverpool as well as Leeds and Bradford in the summer. There were also outward enticements, a very popular evening excursion to Morecambe including admission to the Winter Gardens as part of the return fare.

Gradual decline after the Second World War became more rapid in the 1960s, when many services were carrying fewer than ten passengers in the winter. By 1968 there were no through services beyond Preston apart from one London train that lasted another two years. There was strong local opposition when the branch was reduced to a long single-track siding in 1973 and seven years later the end seemed nigh when it was announced that most of Windermere station would become a supermarket. Facilities on a short remaining platform would be limited to little more than a prefabricated booking office. It was enough to stimulate the formation of a Lakes Line Action Group, which pressed for a larger station building to a design reflecting its status as a railhead within the National Park. The successful outcome was a new award-winning 'chalet style' terminus with the original train shed retained and extended to become part of a high-class store.

Services were radically improved with the introduction in 1984 of through trains from Windermere to Manchester Airport, thus benefiting local businessmen in one direction and overseas visitors in the other. All went reasonably well for the next thirty years with Lakeland's only remaining branch line, which retained as good a diesel unit service as might be expected for a long siding. Hopes ran high with a surprise announcement in 2014 that the branch was to be electrified, but the government then cancelled the project in July 2017. Part of the official justification, widely seen as a convenient political excuse, was that the same month brought designation of the Lake District as a World Heritage Site. Effrontery even extended to arguing that catenary would spoil a protected landscape!

The steam age on the Windermere branch is captured by Derek Cross in four pages of black & white photographs.

Above

What many of a certain age regard as a 'proper train' with a long rake of compartment carriages is headed by 5MT No. 45376. A Windermere to Liverpool service is climbing out of Kendal on 31st July 1964.

Opposite, above

An express from Windermere to Euston is about to join the main line at Oxenholme on 27th July 1963. Rebuilt 'Patriot' No. 45523 *Bangor* has steam to spare.

An up local service to Oxenholme still boasts some compartment stock on 31st August 1963. Stanier 2-6-4T No. 42613 is in pleasantly rural surroundings on the approach to Kendal.

Above

A suprisingly heavy goods working to Carnforth requires banking assistance up the 1 in 80 gradient from Kendal to Oxenholme on 20th July 1962. The train engine is 5MT No. 44874.

Below

About to head downgrade tender-first into Kendal on 27th July 1963 is 4F No. 44440. The decidedly mixed range of wagons and vans has come from Carnforth.

Above

The main line over Shap and its branches retained steam haulage right up to British Railways' official end of steam on 11th August 1968. Eleven days before the fateful date, 5MT No. 44894 is busy shunting in Windermere goods yard, which closed in April 1969.

Below

There is somehow an atmosphere of pending gloom on the same day with the large locomotive coupled to a solitary mineral wagon. Both have seen better days.

Above

A Pathfinder tour forming part of the Regional North West 'Special Weekend' is near Staveley on 23rd May 1993. It is headed by class 20s Nos. 20075 and 20131. On the rear are class 37s Nos. 37708 and 37801, which had earlier taken the train to Windermere.

Below

Moments later the tour has slightly eased towards Kendal and its passengers are briefly offered a view up the Kentmere valley towards Harter Fell.

Above

Class 156 No. 156427 in Regional Railways livery arrives at Staveley on 22nd December 1994. The station, described in an early guidebook as 'built mainly of blocks of slate cemented into a very durable and very countrified-looking wall', now has a basic 'bus shelter' and seat. The derelict platform on the left highlights how the whole of the Windermere branch was reduced to a single-track siding in 1973.

Below

A through service to Manchester Airport provided by class 156 No. 156427 prepares to leave Windermere on 22nd December 1994. In the background is the Windermere Hotel, opened at the same time as the railway to promote Lake District tourism.

Above

In 1986, Windermere gained an attractive new station outside the trainshed, which was converted into a supermarket. Class 175 No. 175106 is ready to depart for Oxenholme on 13th November 2001.

Below

Looking towards Oxenholme on 7th November 2014 with class 185 No. 185110 in First livery. Although more recent than the class 175, passenger comfort was considered inferior.

From its very beginning in 1846 the Furness Railway had its eyes set on Newby Bridge at the southern tip of Windermere. It was the destination for brave travellers who had crossed Morecambe Bay from Fleetwod to Piel pier before continuing by train and coach for a sail on the lake. With the huge expansion of Barrow the company had the resources to reach Windermere by building a seven-mile branch from Ulverston.

Completion was delayed when it was decided that the existing steamer berth, involving vessels in a risky voyage down the headwaters of the Leven, was not ideal for rail/ship interchange. Instead the branch would be continued to a purpose-built quay at Windermere (Lake Side), the extension being completed in time for passenger services to begin in June 1869. These went beyond the tourist market by providing an all-year-round timetabled sevice to Waterhead at Ambleside with connections from branch line to lake shown in the timetables.

The first part of the branch ran hard by the side of the Leven estuary and offered superb panoramas of green fells and rugged mountains. High tide virtually lapped a platform at Greenodd station, where a shelter was provided with windows on all sides. It enabled passegers to enjoy the view while waiting for the train to arrive. The final stretch of line was densely wooded and exceptionally picturesque, the waters of Windermere seeming almost to touch the wheels of the train and the peak of Gummer's How providing a spectacular backdrop.

Boom time in the Barrow iron and steel industry meant it was high noon for the prosperity of the Furness Railway. The company spared no expense with the station buildings on its new tourist line to the extent of causing mutterings among shareholders about extravagant opulence. It used 'patent white bricks' – they were actually a pale yellow – in a decorative style of Flemish bonding, offset by vitrified purple-black stretchers, the whole effect not exactly harmonising with the rural surroundings.

Lake Side, carefully laid out as a combined railway terminus and steamer pier, was especially ostentatious with a ridge-and-furrow overall roof and a slender pointed tower. A narrow gauge tramway across the front of the main buildings provided coal for the steamers from a siding to the west of the station. Running parallel with the lake and platforms was a long veranda, on top of which a 'Palm Court' restaurant allowed tourists to contemplate their surroundings as they dined. Records show that in eleven weeks in the high season of 1906 it served in excess of 800 breakfasts, 4,300 lunches and 3,500 teas. An orchestra complete with viol and harp provided the final soothing touch.

The steamers on Windermere were an essential part of the enterprise from the outset. Just four days after the branch opening there were great festivities and firing of cannon to mark the launch of a new screw-driven steamer *Swan* with accommodation for 450 passengers. She joined the existing paddle steamers, *Dragon Fly*, *Fire Fly* and *Rothay*, which were taken into direct railway ownership. In 1871 the Furness introduced *Raven*, a barge-like vessel that carried freight to and from the Lake Side railhead and provided a water-borne door-to-door delivery service to the many mansions scattered around the lake shore.

The most famous vessel on Windermere was undoubtedly *Esperance*, the graceful steam yacht owned by the Barrow ironmaster H W Schneider, who on opening of the Lake Side branch bought Belsfield House at Bowness. Local folklore relates how each morning he would leave his mansion preceded at a discreet distance by his butler carrying a heated silver salver. A lavish breakfast was then served on board before the industrialist transferred to a reserved first-class compartment to complete the journey to his office. *Esperance* became immortalised as the houseboat in Arthur Ransome's *Swallows and Amazons* and is now preserved at Windermere Steam Boat Museum.

By the Edwardian years developments on the branch centred on the Midland Railway. It became possible to travel from London St Pancras to Lake Side via Leeds and Carnforth without change of train. The company's timetable also promoted a 'Cheap Fast Train' leaving Leeds at 6.40am and by a combination of rail, steamer and coach enabling Grasmere to be reached at 11.50am nicely in time for lunch.

Enormously popular during this period were the Furness Railway's revived crossings of Morecambe Bay by paddle steamers taking Blackpool holidaymakers from Fleetwood to Barrow. Here the many options for 'coach and steam yacht tours through the English Lakeland' included joining a train to Lake Side. Those with real stamina could indulge in a sail along Windermere to Waterhead and then a coach trip to such destinations as Ullswater reached by a challenging haul over Kirkstone Pass. Excursion traffic remained buoyant and was not confined solely to the summer months. If Windermere froze in the depths of winter, special trains were run to bring in skaters from a wide radius. Perversely, it was the Lake Side branch that suffered the first major setback in the winter months. Mounting losses forced the Furness Railway to withdraw winter sailings on Windermere in 1920 and instead borrow buses to provide a substitute connection to Ambleside.

Use of both the branch and the steamers benefited from a change in leisure habits prior to the dawn of mass motoring when summer Sunday services to the Lake District became extremely popular. One instance was a train leaving Leeds at the ungodly hour of 7.20am to reach Lake Side three hours later. There was ample time for a day on Windermere, as the return journey did not start until 7.25pm and another three hours elapsed before arrival back in Leeds. It was a long day out.

Services after the Second World War were confined to the summer months. Combined tours survived and in the 1950s a return fare of ten shillings would convey Morecambe holidaymakers on the 'Windermere Circular' to Lake Side, thence up the lake to Waterhead and back to Bowness by steamer before returning by train via Oxenholme. By the early 1960s the branch was under threat of closure. Cynics feared the worst when most of the line was relaid with experimental concrete sleepers and British Railways announced its intention to expand the Windermere steamer operations. They were proved right when all passenger services were withdrawn in September 1965.

Moves were made to reopen the line with formation of the Lakeside Railway Society a year later. The company established a locomotive depot at Carnforth, which became Steamtown museum. The original hope that it would provide motive power for the branch was dashed when its southern portion was sequestered for trunk-road improvements. Locomotives and rolling stock were accordingly moved to Haverthwaite in 1970-71 prior to severance of the rail link with the outside world. A Light Railway Order paved the way for opening of the Lakeside & Haverthwaite Railway on 2nd May 1973.

Sadly, most of the original buildings at the terminus had to be demolished due to dry rot, but the tradition of trains running in connection with sailings on the lake has now been maintained for over 150 years. Various changes following nationalisation led to steamers trading under the unlikely name of Sealink before they came back into local ownership.

Above

The Northern Fells rail-tour arrives at Lake Side on 29th May 1960 behind No. 42952. The Stanier Mogul is being detached from the carriages prior to running round for the return trip. This ambitious tour had earlier seen tender-first running from Tebay through Oxenholme at a speed of over 70mph!

Below

Operations at Lake Side on 26th August 1964 – the penultimate year of services on the branch. 5MT No. 44877 is preparing to depart for Ulverston and sister locomotive No. 44730 is taking water in readiness to work a later train. The water tower to its left displays the extravagant style of architecture adopted at the terminus. Further left in green and white livery is a camping coach, while at right the platform is alongside the misty waters of Windermere.

Above

Activity at the intermediate station of Haverthwaite on 26th August 1964 includes 5MT No. 44877 arriving tender-first with a train from Lake Side.

Below

A modeller's dream with a tunnel at each end, Haverthwaite sees 5MT No. 44730 working light engine to Lake Side on the same date. The signalman is poised ready to hand over the token for the single-line ahead.

Above

Metrovick Co-Bo No. D5708 runs round its train at Lake Side on 21st September 1964. The inspector on the platform looks very concerned at this movement. The 'patent white bricks' used for the station walls have assumed an unappetising shade of dirty cream. Most of these buildings were destined to be demolished due to dry rot.

Below

No. D5708 has run round and prepares to take its train down the branch to Ulverston. Visible to the left of the lake is the glass roof of the 'Palm Court' restaurant. In its heyday, diners waiting for a Windermere steamer were served meals to the accompaniment of an orchestra.

Opposite, above

Health & Safety would no doubt have plenty to say today about the general mayhem on 2nd May 1973, when the surviving portion of the branch reopened as the Lakeside & Haverthwaite Railway. Schoolchildren throng the tracks as the inaugural train prepares to depart from Haverthwaite. They are standing in front of *Caliban*, a 0-4-0ST built by Peckett in 1937. It was later transferred to the Ribble Steam Railway at Preston.

Opposite, below

Enthusiasts are even atop the signal gantry as the special departs from Lake Side. It is double-headed by Fairburn 2-6-4 tanks Nos. 2073 and 2085.

Right

Bishop Eric Treacy, who performed the reopening ceremony, gives an interview to the local press.

Above

There is a much calmer scene on 3rd June 1973 as green-liveried 5MT No. 44806 arrives at Lake Side. Named *Magpie* later that year, it had a brief spell on the line before being moved to the Steamport Railway at Southport and then on to the North Yorkshire Moors Railway.

Above

No. 2085 departs from Haverthwaite for Lake Side on 9th September 1976. The 2-6-4T was built at Brighton to an LMS design.

Below

A side elevation of No. 2085 shows its striking livery of Caledonian Railway blue, lined in black and white, with deep red cylinders and frames.

The Furness Railway branch to Lake Side was preceded by an earlier line that was not nearly as successful. Its origins were the famous copper mines high on the fells above Coniston which by the mid-1850s were the most productive in Britain. The big problem was having to transport the ore down from the mines by carts followed by barges on Coniston Water and then more carts before it could be shipped out by sea.

Hence the promotion by mining interests of an eight-mile branch extending from Coniston down to the Furness Railway at Broughton. It was in trouble almost from the outset. Miners of the time long cherished what they called 'the elusive dream'. Rather like winning today's lottery, it was the eternal hope that untold wealth was just around the corner but so often it was the exact oppposite. Such was the case at Coniston where opening of the branch in 1859 from a new junction station at Foxfield coincided with what proved to be peak output at the mines. As this ebbed away, there were growing doubts about a railway conceived primarily to carry mineral traffic. It became clear that it was in the wrong place at the wrong time sited away from both the central Lake District and civilisation. Moreover, its Coniston terminus was high above the village and lake in order to handle ore traffic but was far from ideal for passengers.

Tourism in the Lake District was rapidly expanding and had the branch been designed with this source of traffic in mind it would almost certainly have diverged from the Carnforth-Barrow route near Ulverston. It would then have followed the Crake Valley due north to run alongside Coniston Water before terminating in the centre of the village rather than at the top of a steep access road.

The Furness Railway took over the Coniston branch in 1862 and did its best. A group of directors commissioned the famous *Gondola* steam yacht to operate seasonal sailings on the lake. From 1870 it formed an integral part of 'Circular Tours' from Barrow using a combination of rail, horse haulage and steamer. After reaching Coniston by rail, passengers were carefully taken down the hill by omnibus, offered relaxation on the *Gondola* as far as the foot of the lake and returned to Barrow after further horse-hauled conveyances took them to Greenodd station. A variation included two lakes with the tour going first to Lake Side prior to a sail up Windermere to Ambleside and then road travel to Coniston Water.

Viewing this activity with distinctly mixed feelings would be the distinguished if at times deranged conservationist John Ruskin. Seeking a haven from what he regarded as the worst evils of the industrial revolution, he moved to Brantwood overlooking the eastern shore of Coniston Water in 1871. He made no secret of the fact that he detested anything to do with railways and in September 1874 penned a typically bombastic letter to a local paper. It condemned the evils of a roundabout train journey to Ulverston with its perils of getting drunk at stations along the way compared with a healthy walk through pastoral scenery.

The Coniston terminus of the branch was an especially attractive building, gradually extended in a Swiss cottage style of architecture until it had an overall roof with an arched end-screen springing from the platforms, gabled drystone walls and a large tea pavilion. It had well-tended gardens and was in an unrivalled setting dwarfed by the magnificent backdrop of the Above Beck Fells with Coniston Water down in the valley bottom.

The copper mines closed in 1889 but every effort was made in the Edwardian years to develop the existing 'Circular Tours' in the same way that changed the fortunes of the

Lake Side branch. On several of these tours the passengers journeyed back to Barrow in what timetables grandiosely titled the Fleetwood Boat Train before returning across Morecambe Bay to their holiday accommodation. Few branch lines can ever have carried a boat train in quite the same way!

Even though the cross-Bay steamers were not reinstated after the war, Blackpool holidaymakers could still head this way. The LMS developed a through service, which most years left the resort at 8.50am and reached Coniston at noon. The return working departed at 6.0pm. It surely cannot have been weariness of the attractions of Blackpool's 'Golden Mile', but today it seems strange that there was a demand for a day's escape involving over six hours of train travel for an afternoon on Coniston Water. It was almost on a par with Glaswegians heading for Keswick as described

in chapter 1. There must have been many a breathless climb up the long hill to the station by passengers anxious to catch the only train that would take them back.

More surprising is that the service survived the Second World War, running on Tuesdays and Thursdays only until the end of the 1958 summer season. This was a significant year as it had become increasingly apparent that the branch continued to be hamstrung by its inherent drawbacks first arising a hundred years earlier. An attempt in 1954 to replace steam-hauled push-pull local services by an experimental three-car diesel set proved unsuccessful. Large annual losses were now being incurred and the branch became the first Lake District line to lose its passenger services in October 1958. Complete closure took place in April 1962.

Opposite
Two views from different perspectives of Coniston station in its superlative setting. Taken on 16th February 1958 – the last year of passenger operations – there is significantly not a sign of life. Coniston copper mines, which brought the branch into existence, were sited in the fold among the mountains visible at left in the lower photo.

In the same way as the more modest Coniston branch, it was minerals and not tourism that created the only railway to run through rather than merely into the Lake District. It was spearheaded in Whitehaven, where there was one big problem in making the most of hematite ore deposits discovered in the nearby fells. A means had not then been perfected for converting coal from the surrounding mines into coke, which was having to be brought from County Durham by an indirect route through Newcastle, Carlisle and Maryport.

A similar problem at Barrow had led to the formation of a line climbing over the Pennines from South Durham through to Tebay, from where coke could be taken over the West Coast main line and thence into Furness. A branch from this cross-country route was under construction from Kirkby Stephen to Penrith and hence it became feasible to build a 30-mile link through Keswick to Cockermouth, which was already connected with the main coast line at Workington. It was backed by the London & Northern Railway, which used it to gain access to the lucrative traffic centred on Whitehaven and its hematite. The engineer was Thomas Bouch, whose precipitous downfall on collapse of the first Tay Bridge was then inconceivable.

Mineral traffic came first and began in October 1864. Today it seems extraordinary that a single set of enginemen would leave Shildon around 4.0am for the formidable climb over Stainmore to Penrith. There was then a second slog up to a summit of 889ft at Troutbeck, almost as high as Shap, before finally handing over to another locomtive at Cockermouth. This was only half the story, as a quick turn round was required before heading all the way back again to complete a gruelling 180-mile round trip lasting at least sixteen hours. Apart from carrying coke on a more direct route, hematite ore was taken eastwards to iron and steel works on Teeside.

Although seldom heralded as such, the line provided one of the outstanding scenic railway journeys in England when opened to passengers in 1865. After the Troutbeck summit there were splendid views of the striking peak of Saddleback, especially from the twelve-arch Mosedale viaduct. The character of the route now changed completely as it plunged into the Greta gorge. From above there was little indication of a railway in the bottom apart from eight slender box-string girder bridges criss-crossing the river. Emerging suddenly at Keswick, the floodplain of the Derwent was crossed before the line swung northwards and hugged the western shore of Bassenthwaite Lake. This final spectacular stretch with broad panoramas of Skiddaw preceded more gentle surroundings on the approach to Cockermouth.

It was not long before carrying tourist traffic as well as minerals entered the equation. The superb situation of Keswick in almost Alpine style among high mountains soon attracted more visitors wanting to share the uplifting powers of the secenery as earlier publicised by Wordsworth.

The railway duly provided a first-class hotel next to the station with its own covered entrance from the platform. Described as a 'substantial pile', it operated combined road and rail tours through the northern half of the Lake District.

As with other lines such as the Furness, passenger revenue became increasingly important as mineral traffic gradually declined. The Edwardian years saw a heyday of through services to Keswick from a wide range of destinations including London, Manchester, Liverpool, York and Newcastle. The Midland Railway introduced the Lake District Express, which left Leeds at 10.0am and after running non-stop to Penrith via Ingleton enabled passengers to reach Keswick at 12.50pm in time for lunch – something that would rarely be possible more than a century later on today's roads. There was a traffic surge each July when those seeking closer communion with God among the mountains flocked to the Keswick Convention. Numerous special trains to this evangelical conference included 'The Budd', a through express from Euston named after the London businessman who promoted it. Well-loaded and operating on a fast daytime schedule to Penrith, it was then normally double-headed through to Keswick.

Many of those attending saw themselves as 'missionaries to the heathen', which may also have been the view taken by one shareholder in reference to local train services. At a stormy meeting he quoted from Genesis, 'God hath made everything that creepeth upon the Earth', to support his claim that the line was the only railway to be mentioned in the Bible. Moreover, its 'dreadful rolling stock' was antiquated, uncomfortable, dirty and much of it was the 'refuse' of the London & North Western Railway.

Matters gradually improved, but it was not until strengthening of bridges in the mid-1930s that locomotives of any size could reach Keswick. A different challenge arose during the Second World War when the Keswick Hotel was used by the prestigious Roedean School and its girls following evacuation from a vulnerable position near Brighton. End-of-term departures to Penrith and on to London were heavy double-headed trains leaving at the ungodly hour of 5.30am. This was apparently no problem for the formidable headmistress, Dame Emmeline Tanner, who regularly organised impromptu expeditions to the summit of Helvellyn and was long remembered

locally for her awesome oration on 'Two Gels seen eating Chips in Keswick'! She even put fear into the higher echelons of the LMS with instructions being issued to railway staff that the Roedean trains should take priority over all other wartime traffic.

Fundamental change was then conspicuous by its absence until 1955 when some of the first DMU train sets in Britain took over augmented and accelerated services from Penrith to Workington. They failed to save the line and the section west of Keswick was closed in April 1966. An annual highlight continued to be 'The Budd', which was still a heavy train serving the Keswick Convention and steam-worked until as late as 1967. The following year it had to be replaced by a shuttle service from Penrith, as the remaining part of the line became an extended siding with stations reduced to unstaffed halts. All that was left was a two-car DMU, little used in winter and grossly overcrowded in summer, which lingered on until it ran for the last time in March 1972.

It was not quite the end of a sad saga. By this time controversy had erupted over proposals for a new A66 trunk road from Penrith to Cockermouth, closely paralleling the railway and using part of its trackbed alongside Bassenthwaite Lake. A protest meeting on Latrigg summit, near Keswick, concluded with the evocative words: 'In the future someone will realise that the railway line that once ran so unobtrusively below us could have carried all those people and all those heavy loads at a tiny fraction of the cost and damage to the environment.' These were prophetic words but the road went ahead and it was too late to save the railway. It was at least the last line serving the Lake District to be closed to passengers.

Seven pages of black-and-white photographs by Derek Cross portray traffic from Penrith west to Cockermouth in the early 1960s.

Opposite, above

A through service from Keswick to Manchester comes alongside the West Coast main line at the approach to Penrith on 18th August 1962. About to pass the large signal box controlling the junction, it is headed by Ivatt 2MT No. 46491 – a class closely associated with such workings in their final years.

Opposite, below

The same location on 31st July 1964 sees another 2MT No. 46434 bringing lime from Flusco Quarry, four miles up the branch at Blencow. The direct connection with the main line was removed four months later at the end of November.

Above

The way in which the Penrith to Keswick track immediately climbs away from the main line at 1 in 79 is clearly evident in this photo taken on 27th July 1963. The mere five coaches behind 'Princess Coronation' No. 46229 *Duchess of Hamilton* immediately show that this is the down Lakes Express.

The photos on these two pages provide further views of the Lakes Express.

Above

It is a tank engine – Stanier 2-6-4T No. 42571 – that is taking the service south from Penrith to Oxenholme in July 1962. Waiting at the impressive lower-quadrant junction signal is 2MT No. 46434 with just a single wagon from Flusco Quarry. Such workings were definitely not economic!

Below

The length of the northern portion of the Lakes Express was far from fixed. On 22nd July 1960 it extends to seven coaches behind Stanier 2-6-4T No. 42594. It is leaving Penruddock, close to the 889ft summit at Troutbeck, and starting to drift downgrade for the next eight miles into Penrith.

Above

The down service arrives at Keswick on 21ˢᵗ August 1964 behind 2MT No. 46434. On the left is the spacious island platform providing ample accommodation for tourist traffic, promoted on a selective basis in order to preserve the sedate image of a town that had no desire to become another Bowness or Windermere.

Below

The same locomotive prepares to depart from Cockermouth the previous day with a three-coach up Lakes Express. Attractive glazing on the waiting area provides practical protection in a part of the world where the weather can often turn hostile.

Opposite, above

A special from Keswick to Carlisle for a sports fixture approaches Troutbeck on 31st July 1965. The 2MT is No. 46491.

Opposite, below

2MTs Nos. 46458 and 46455 bring the empty stock into Keswick on 17th July 1965 for 'The Budd' – a Convention special returning to London. Carefully cleaned locomotives became a notable feature of such workings in the final years of steam on the Keswick line.

Above

Having run round, the same pair of 2MTs are coming to the end of a long slog in lifting the heavy Convention special out of Keswick over an eight-mile climb mainly at 1 in 62. Steam is about to be shut off on the approach to Troutbeck summit. Running 'wrong line' reflects future singling following imminent closure of the line west of Keswick.

Two final photographs by Derek Cross capture the setting of the sorely missed and only railway to run through the Lake District.

Above

The 2,847ft peak of Saddleback with its unmistakable shape forms a superb backdrop. 2MT No. 46491 is near Troutbeck with a Keswick to Crewe service on 31st July 1965.

Below

The Skiddaw range has a light covering of snow on 2nd April 1966. A DMU is arriving at Keswick with a service from Workington to Penrith two weeks before the cessation of such workings.

Above

Another instance of a pair of well-groomed 2MTs on 13th June 1964. Nos. 46426 and 46458 are running through Bassenthwaite Lake station towards Keswick with the Solway Ranger rail-tour on 13th June 1964. In such a location it is scarcely surprising that the two camping coaches on the right were extremely popular.

Below

The Solway Ranger has passed through Keswick and is beginning the long climb to Troutbeck.

Sad tidings at Keswick station on 4th March 1972 – the last day of passenger services on the branch from Penrith. An extra long DMU has been provided instead of the normal two cars. On the left is the covered awning built in a more comfortable age to provide direct access from the platform to the Keswick Hotel.

Below

The buffer stop and the last train symbolise the end of the line. The state of the island platform is a sorry contrast with the photo on page 111. On the right the station buildings are dwarfed by the massive hotel.

4

THE NARROW GAUGE

RAVENGLASS & ESKDALE RAILWAY

Closure of the Penrith to Keswick line in 1972 was not the absolute end of Lake District railways in magnificent mountain scenery. There is a rare survivor that should never be overlooked. The Ravenglass & Eskdale did well to escape objections from pioneer conservationists who had much to say about proposed lines in Borrowdale and Ennerdale. For those opposed to any change, it had the definite drawback when proposed in 1873 that it was to use steam locomotives. Above all, it was to terminate within five miles of Scafell Pike – highest of all the Lake District summits at 3,210ft. Its saving grace may have been that it was a three-foot narrow gauge line, which in the prevailing attitude of the time was not always seen as a proper railway.

It was the discovery of hematite ore close to the village of Boot that led to formation of the Ravenglass & Eskdale. By 1875 it was carrying the ore down to the Barrow–Whitehaven line, where it was transshipped into standard gauge wagons. There were immediate problems when it was found to be so badly built that a Board of Trade inspector refused to sanction its opening to passengers for another eighteen months. Services finally began in November 1876. It was the last railway to be built in the Lake District.

There was already financial trouble. The company was unable to pay the contractor and within six months became subject to an Order in Chancery followed by appointment of a receiver. In 1882 it lost its main source of traffic when the mines failed, periodic attempts

at revival being unsuccessful. The line was left to eek out a living by conveying tourists in its three carriages and assorted mineral wagons, endeavouring to attract custom by fancifully advertising itself as 'the route to the Alps'.

It was soon extremely run down, as captured in 1903 by local author Mary C. Fair. She stated that the train 'lurches, and groans, and rolls along – you speculate whether the bottom will fall out of the carriage, the train pull up the rails, or the whole affair topple into the river'. Closure to all traffic took place in November 1908 but occasional goods trains were reinstated until April 1913 when operations finally ceased.

The First World War brought an unexpected revival that was nothing less than total metamorphosis. Early in 1915 the derelict remains were inspected by a group of enthusiasts who included W.J. Bassett-Lowke, famous for his role in creating the model railway hobby. It was an inauspicious time with trench warfare raging in Flanders, but they nevertheless decided to create a 15-inch gauge miniature railway by leasing the moribund line. It was a move that received high praise, although a less charitable view is that they were keen to provide themselves with a reserved occupation to avoid serving in the trenches.

Services from Ravenglass to Muncaster began in August 1915 and the line was progressively extended to reach Boot two years later. Some of the early miniature locomotives were winded by the final 1 in 37 climb and a new terminus

Above

With the Ravenglass & Eskdale threatened with sudden demise, what could well have been a farewell rail-tour from Leeds was organised by the Railway Correspondence & Travel Society [RCTS] on 4th September 1960. With its fate still unknown, the gloomy mood was not helped when a poorly performing unrebuilt 'Patriot' No. 45503 *The Royal Leicestershire Regiment* arrived at Ravenglass about an hour late. Spirits were later raised when an immaculate pair of Ivatt 2MTs took the special from Workington to Penrith, and there was a stunning performance when preserved Midland Compound touched over 80mph when heading back through Settle.

was opened at Dalegarth in May 1920. There was a revival of combined rail/road tours in western Lakeland, this time by charabanc, which took participants to see the wonders of Wastwater before they rejoined the train at the intermediate station of Irton Road.

A committed benefactor was Sir Aubrey Brocklebank of nearby Irton Hall. Head of the shipbuilding company that later controlled Cunard, he saw the need to ease dependence on seasonal traffic and supported the 1922 reopening of a granite quarry at Beckfoot, close to Dalegarth. Two years later a crushing plant was built at Murthwaite and then in 1928 it was decided to eliminate transshipment problems and connect it to Ravenglass by a 2½-mile standard gauge line. The 15-inch and standard gauge tracks were interlaced on the same sleepers.

The entire enterprise was acquired in 1949 by the Keswick Granite Company, which a decade later decided to auction the line to the highest bidder. The arrival of scrap merchants seemed imminent and a rail-tour was arranged on 4th September 1960 for what could well have been a final wake. Three days later there was a happier outcome when a newly formed preservation society was able to make a successful bid at a public auction. Half the purchase price was met by private individuals, who held the line and its historic 15-inch gauge locomotives in the highest regard.

It would once have seemed inconceivable that the railway affectionately known as La'al Ratty would outlive standard gauge lines such as the Coniston branch. Yet it has since moved ever forwards to become a major Lake District tourist attraction and feature high on the list of 'must visit' lines for countless enthusiasts.

The photographs on the next four pages capture the atmosphere of the rail-tour on 4th September 1960.

Above
Complete with RCTS headboard, *River Esk* prepares to depart from Ravenglass. Designed by the famous engineer Henry Greenly, the 2-8-2 was built in 1923 by Davey, Paxman & Co.

Below
The Muir-Hill petrol tractor, built in 1928, shunts open carriages at Ravenglass.

Opposite, above

More than half-way along the seven-mile line, *River Esk* pauses at Irton Road. The author was a lanky teenager with sticking-out ears when participating in this memorable day. Exceptionally tall, he is convinced that he is the figure standing in the first open carriage!

Opposite, below

The Muir-Hill petrol tractor brings open carriages into the line's terminus at Dalegarth.

Above

Originally built in 1894, the railway's oldest locomotive 0-8-2 *River Irt* approaches Dalegarth. It is passing the terrace cottages originally provided for local miners and their families.

Above

Passengers return to Ravenglass on 4th September 1960 to rejoin the rail-tour back to Leeds, uncertain what fate was in store when the railway was put up for auction three days later. The word that it had been saved soon spread!

Opposite, above

A more assured era on 8th June 1976 as *River Mite* passes Muncaster Mill on its way into Ravenglass. This 2-8-2 was built by Clarksons at York in 1964-66.

Opposite, below

The same day and the same location with another train headed by *River Esk*. Its livery has been changed from green to black to distinguish it from the near-identical *River Mite*.

More than fifty years have passed since closure of the Penrith to Keswick line, but it is still possible to see locomotives at work close to its former trackbed. Moreover, these will be in steam and can be enjoyed in a location offering glorious views of Saddleback and the Skiddaw range of mountains. The enticing location is Threlkeld Quarry & Mining Museum, four miles east of Keswick.

The granite quarry, originally opened in 1870 using a narrow gauge line to extract its output, provided railway ballast and later supplied stone for the Thirlmere reservoir scheme. It finally closed in 1982 to become a museum with a two-foot gauge railway that has some steep gradients including a section of 1 in 20. Vintage machinery and cranes provide an authentic industrial atmosphere and the museum has gained a growing reputation for its photo opportunities. Galas with visiting locomotives have become a 'must-attend' event for many enthusiasts.

Resident locomotive is 0-4-0ST *Sir Tom*, which arrived at Threlkeld in 2001 for complete restoration. Built by Bagnall of Stafford in 1926, it is named after Sir Tom Callender of British Insulated Callender Cables – the Kent concern where it spent its entire working life until retirement in 1968. Today it hauls passenger trains within the quarry.

Among the wide range of visiting locomotives has been 0-4-0 well-tank *Utrillas*. It derives its name from the Minas de Utrillas near Zaragoza in Spain, where it was sent to work following construction by Orenstein & Koppel in 1907. After a period on the West Lancashire Light Railway it moved to the Lancashire Mining Museum at Astley Green, close to Manchester.

More distinctive and attracting a great deal of interest was the vertical-boilered *Paddy*, built at Wilbrighton Wagon Works in 2007 using older components. Now at the Statfold Barn Railway, it has been renamed *Howard*.

Opposite

Resident locomotive *Sir Tom* in charge of four flat wagons on 26th April 2007. Threlkeld village on the far side of the valley is clearly visible in the lower photo.

Visitors to Threlkeld Quarry on 27th July 2012.

Above

Well-tank *Utrillas* with its train of 'V-skip' wagons. The shed survives from the quarry line opened in the 1870s.

Below

The slopes of both Skiddaw and Saddleback are in the background as *Utrillas* faces a steep gradient to the upper reaches of the quarry.

Above

Polished brass and plenty of steam are evident on vertical-boilered *Paddy*, then only five years old but looking much more of a veteran.

Below

An excavator of suitable age helps to capture a bygone era of steam-worked industrial railways. *Utrillas* is on the left and *Sir Tom* is beyond the 'V-skips'.

FURTHER READING

Succumbing to vanity, I begin with my first book. It had the same title as chapter 1 in the present book, *Main Line Over Shap*, and was published way back in 1967. A year later came *Cumbrian Coast Railways* on the great way round from Carnforth to Carlisle via Barrow and Whitehaven. Research then got more serious, eventually resulting in 1983 in *The Lake Counties*, Volume 14 in *A Regional History of the Railways of Great Britain*, a series widely accepted to have stood the test of time.

In writing *Railways and the Lake District*, I have relied heavily on a series of lengthy contributions to *Backtrack* – 'Britain's Leading Historical Railway Journal'. These are: 'Penrith – A Bygone Junction' (October 2016); 'Two Dukes and a Lord – The Nobility and the Railways of Barrow' (May 2018); 'One Lord and no Dukes – The Nobility and the Railways of Whitehaven' (October 2018); 'Rails to Windermere' (June 2020); 'Rails in Western Lakeland' (November 2020).

My 2017 book for the Cumbrian Railways Association [CRA], *An Introduction to Cumbrian Railways*, embraces the Lake District. Excellent works in the CRA 'Railway Histories Series' include *The Kendal & Windermere Railway* by Dick Smith (2002), *The Ulverstone & Lancaster Railway* by Leslie R. Gilpin (2008), and *The Coniston Railway* by Michael Andrews & Geoff Holme (2nd edition, 2012).

Major studies published locally by Barrai Books are *The Furness Railway: A History* by Michael Andrews (2012) and *The Railways of Carnforth* by Philip Grosse. Also important are *Rails through Lakeland*, Harold D. Bowtell's illustrated history of the Keswick line (1989), and *Ravenglass & Eskdale Railway* by W.J.K. Davies (new edition, 2000).

Last but definitely not least, words are only a small part of the present book. For more of Gavin's superb photos, I especially commend *Steam Photographer: My Favourite Pictures of the BR Era* (2021) with more than 400 outstanding images of steam from the period between 1959 and 1968.